How to Select & Install Your Own Speakers

No. 1034
$9.95

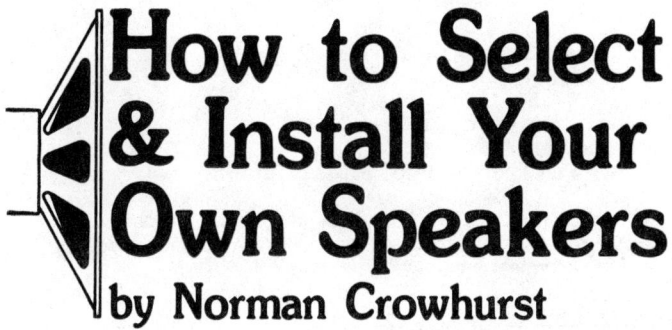

How to Select & Install Your Own Speakers
by Norman Crowhurst

BLUE RIDGE SUMMIT, PA. 17214

FIRST EDITION

FIRST PRINTING—FEBRUARY 1979

Copyright © 1979 by TAB BOOKS

Printed in the United States of America

Reproduction or publication of the content in any manner, without express permission of the publisher, is prohibited. No liability is assumed with respect to the use of the information herein.

Library of Congress Cataloging in Publication Data

Crowhurst, Norman H.
    How to select and install your own speakers.

    Includes index.
    1. Loud-speakers—Amateurs' manuals. I. Title.
TK9968.C76   621.38'0282   78-20978
ISBN 0-8306-9823-X
ISBN 0-8306-1034-0 pbk.

Cover photo courtesy of Herald Electronics.

Preface

That this book is needed, is shown by the number of letters received on the subject. These letters often want answers without fully asking the question, which means the inquirer does not know enough about the principles of sound reproduction to be able to ask what it is he needs to know.

This book is organized in a form that overcomes this difficulty. It does this by considering variables that cause problems. This gives a basis for making each individual choice. Some will want to know what ready-made loudspeakers to buy and where to place them. But one needs to know what kind of room these people have and what they like to listen to before such advice is possible.

Others are more ambitious, wanting to build their own, or perhaps to make their system built-in. In this situation, doing it wrong can be disastrous and much harder to correct. Whichever you want to do, it is all in this book. Take it one step at a time, and you will get the results you want.

All kinds of equipment are available. But you need to know how to put it together so it works in the way you expect when finished. By doing it right, you will be assured of countless enjoyable hours listening to whatever it may be that gives you pleasure.

Happy listening!

Norman H. Crowhurst
Dallas, Oregon

Other TAB books by the author:

No. 110 *Electronic Design Charts*
No. 494 *Audio Systems Handbook*
No. 513 *Understanding Solid-State Circuits*
No. 546 *Electronic Musical Instruments*
No. 588 *Basic Electronics Course*
No. 634 *Basic Audio Systems*

Contents

1 Where Do You Want It and Why? 7
Reasons for Differences—Why the Differences—How Do You Tell the Difference?

2 What Your Loudspeaker Has To Do 15
Frequency—Wavelength—Range—Kind of Sound Distribution—Listening Directively—Loudspeaker Directivity—Mono, Stereo, Quad—Binaural—Early Stereo—Stereo Deficiencies—Efficiency—Power.

3 Fitting a Loudspeaker To Its Job 31
The Power of Selective Listening—Achieving the Objective—Room Quality—Room Size—The Car as a Room—Physical Fit.

4 How Each Kind Does Its Job 43
Moving Iron—Moving Coil—Baffles—Horns—Practical Baffles—The Infinite Baffle—Effect of Resonance—Box Size—Bass Reflex—Low-Frequency Differences—High-Frequency Differences—Multiway Systems—Efficiency—Acoustic Suspension—Loaded Reflexes—The Flat Radiator.

5 Picking The Right Kind For The Job 65
Handling the Frequencies—Directivity Questions—Single Unit or Multiway?—Integral Sound—Crossovers—Different Kinds of Directional Effect—Relative Size—Difficult Rooms—Omnidirectionals.

6 Ways To Install Simple Inexpensive Units 83
Round or Oval—Getting the Bass—The Bottom End—Built-In Resonators—Home-Built Horns—Horn Design.

7 Ways To Install Multiple Simple Units 101
One Way of Getting It Together—Columns.

8 Putting Together More Sophisticated Types 111
Progress in Loudspeaker Design—The Problem—Acoustic Suspension—Loaded Reflex—Designing Your Own.

9 Things That Can Ruin Your Performance 131
Placement Problems—Importance of Sealing—The Right Unit for the Box—Duct Design—Effective Box Size—Matching the Room—Cars—Snug Mounting.

10 The Importance Of A Thing Called Phase 145
What Is Phase?—Is Phase Important?—Checking for Phase—System Phasing—Relations Between Stereo and Quad—Crossover Phase—Checking Crossover Phase—How Important Is It?

11 Feeding Loudspeakers Correctly 161
Matching—Multispeaker Systems—Finer Adjustment—The Figuring—Damping—How Much Power—Extension in Another Room—Phasing Again—Multiway—Choice of Crossovers—Electrical or Electronic Crossover—Feeding the Full-Beam Multiple Unit—Constant-Voltage Line Distribution.

12 Making Good Connections 197
Connecting Wire—Making Connections—Soldered Joints—Extra Connections—Wiring—Routing Leads—Fault Tracing—Cables To Use.

Glossary 211

Index 231

1

Where Do You Want It And Why?

Most people think of a loudspeaker as being simply a loudspeaker. It can be large or small, good or not so good, but its purpose is to reproduce the sound that you want to hear—right? It is not quite that simple. We have in our files many letters from people who thought it but who found out otherwise the hard way.

In some instances, the letter writer had listened to a good high fidelity system in someone else's home. Then, when asked by his church to provide sound reinforcement for the congregation to hear the minister better, he ordered these same loudspeakers to install in the church. But they did not work out so well in church.

In other instances, the letter writer had heard a good installation in a theater, found out what kind of loudspeakers the theater used, and bought the same kind for his home high fidelity system. And when he got them installed, they did not sound as good as they did in the theater.

We could extend the list of these mistakes that people make, and there could be many reasons for such results. Even more puzzled is a letter writer who heard a good stereo system in someone's home, went out and bought one like it for his own home, and discovered it sounded awful. The letter writer's first conclusion is that he must have purchased a "lemon."

In some instances where this has happened, he has gone to the trouble of swapping the one he bought with the good one he heard in his friend's house. What they then find is that either of them sounds equally good in the friend's house, while both of them sound equally bad in his. The fact is, both units were good, but the particular system does not suit both rooms equally well.

A particularly important significance is to be derived from observations like this. Nobody can honestly claim such a thing as a "best" loudspeaker without qualifying the statement as to what it is to be used for and where it is to be used. If you read any such blanket claims, we strongly suggest that you not believe them!

REASONS FOR DIFFERENCES

To help you understand what can affect your choice of loudspeaker (other than cost), we need to take a look at the factors involved. Just what are they?

We can divide these into two parts: where you want to do your listening and what it is you want to listen to.

Where You Listen

The obvious difference that depends on where you listen is the size of the room. It could be a living room, a theater, a church, or perhaps even a large amphitheater or stadium. In this book, we will assume that our readers will not be installing a stadium system, but it may sometimes be pertinent to discuss such a system for comparison. Most often the listening area will be a living room or some other room of various size in your house. Or it could be in your car, which is a lot smaller.

That relates to size, but size is not the only difference. Some of the letters we received came from people saying that the two rooms—one where the system sounded good, one where it sounded bad—were the same size and some of them were even the same shape. So what else affects how a system sounds in a room?

Furnishings affect the quality of the room. By this we mean whether it is "dead," "live," or somewhere in between. Probably the best way to check this property of a room is to clap your hands in it. How does the room respond to your handclap?

In rooms that are well furnished, with heavy carpet on the floor, plenty of well-stuffed furniture, heavy drapes, perhaps an acoustic-tiled ceiling, the handclap will sound quite dead. You will hear the clap but with absolutely no echo. In fact, you may have to clap fairly hard to make a sound that is reasonably audible. That is a dead room.

In rooms that are sparsely furnished, with tiled floor, plaster ceiling and walls, and furniture that has little cushioning, the handclap will come back to you crisp and often repeated. The effect is very different from the dead room. And each repetition or echo of the clap sound may have quite an edge to it. This is called a live room.

Now perhaps you begin to see why loudspeaker systems can sound different in different rooms. Suppose you have a system that sounds good in a dead room. You put it in the live room, and the sound you hear is confused, jumbled, and perhaps, even sounds distorted. In fact, you would say it does not sound like the same loudspeaker at all.

On the other hand you have a loudspeaker system that sounds well in a live room. You put it in the dead room and it sounds inadequate, mushy, perhaps even distorted in a different way. Again, it does not sound like the same loudspeaker.

You need to be aware of this kind of difference and why it happens if you are to pick a system that will sound well in the room where you want to use it. As you progress through this book, we will give you some steps to obtain the sound you want. The first step is to know what differences to look for.

What You Listen To

The other major basis in choosing a loudspeaker is what you want to listen to. Perhaps the most obvious division in this department is between speech and music. You may listen to a tape recorder, CB radio, or something else that is commonly used for speech and think to yourself, "Isn't that clear?" Clarity is certainly what you need in listening to speech.

And don't you want to hear music clearly, as well? Of course. But what constitutes clarity in speech is very different from what constitutes clarity in music. For example, a tape recorder may record speeches with beautiful clarity, but when you try it on music, it whines like an alley cat. The music sounds horrible.

That would be due to poor speed constancy in the tape drive, which admittedly has nothing to do with the loudspeaker used. But it illustrates the different requirements for speech and music, and there are differences that do show up in the loudspeaker. For example, what sounds brilliantly clear on speech can sound raspy and broken on music.

On the other hand, what gives beautiful reproduction of music can sound like someone talking through a wet balnket with speech. This is another variable you need to be aware of when thinking about a loudspeaker installation. After learning more about how loudspeakers work, knowing what makes a loudspeaker sound good under each of these circumstances will be obvious. To begin with, you need to know the causes of these subtle variables, or at least some of them.

WHY THE DIFFERENCES

An old piano tuner, who seemed to be stone deaf when I tried to talk to him, once explained to me that a different part of the human ear is used for listening to music from that used for listening to speech. That would explain why he could tune the piano very effectively but could not hear what I said. But a surgeon who knows about the structure of the human ear might argue with his explanation.

Whether or not his explanation was correct, his statement was based on very real observation. There are reasons for this. The ear is a complicated mechanism, and while the same elements are used for all hearing, they are used a little differently for listening to various sounds. It would be difficult to argue with the fact that my tuner friend did his job very well, in spite of the fact that he had difficulty hearing what I said.

When you encounter a situation like this, it is helpful to think about how it came about. Presumably an older man, who has been tuning pianos all his life, was not originally deaf to voices, so he could not hear what the owners of the pianos said, when he talked with them. That must have been a condition that came on gradually.

You will realize that a man who does a job like that uses his hearing with much concentration. He has to listen closely to the musical notes he is adjusting to be sure when they are right. So he develops a "manner" of hearing. Can you imagine that? Think what it means, and how his hearing might develop as a result of doing that.

Now, having got the picture thus far, what happens when he starts losing his hearing? He is going to hang on tenuously to the concentration he has always used, isn't he? What people say is less important to him. So gradually, he gets to hear one, but not the other. At first, perhaps, he is unaware of the difference, and those he meets with are unaware also. But as his hearing gets worse, the difference becomes more marked.

So think about what distinguishes speech and music. In speech, you do not identify tones, or notes, as such. A person can speak in a high-pitched or low-pitched voice, and some voices may be almost musical to listen to, but speech does not have a definite pitch to it, like a piano note does, for example.

Then you can listen to all kinds of musical instruments play the same musical note and, if you have listened critically to music before, you can tell what instrument is playing that note. Each has its own characteristic sound, as well as being able to play various notes on the musical scale.

There is a similar, but different, distinctive capability about listening to speech. You can identify who is talking by the characteristics of his or her voice, what nowadays voice analysts call a voice print. Isn't that the same kind of thing as the differences between various musical instruments?

It may be similar, but it is definitely not the same. The fact is, that what you listen for, to detect the differences, is not quite the same kind of thing in each case. You will find it instructive and helpful to think about that difference, from your own listening experience.

An electronically obtained voice print is a pretty complicated display of the person's speech, in frequency content, relative intensity, and so forth, measured against time; the prints show, visually, the sound of various words the person says. You could see another kind of voice print on an oscilloscope, as a squiggly line, or perhaps, through a magnifying glass or microscope, looking at the groove of a phonograph recording of the same person's voice.

Those two kinds of voice prints are quite different ways of showing the sound in what someone says. The voice print gives a better graphical display, from which to show that two people's voices are different. It would be difficult to recognize very much from the oscilloscope trace, or the phonograph groove magnification.

But now, as you listen to someone talk, is the way you recognize what they say, as well as who it is speaking, like either of those visual methods of examining the spoken word? What is it you listen for? Perhaps one thing might be the peculiar twist a person gives to the pronunciation of certain letters, like r—a way of rolling it, or perhaps a musical lilt that goes with it. Or it may be a different way of pronouncing the vowel sounds.

It is something to which you can attune your hearing, so you can pick up what that person says, as distinct from other conversation that may be going on, about the same time. It could relate to the difference you would see on voice prints, but even a voice print would not find it easy to correlate the two.

How do you do this? You probably do it, without ever thinking about it. We tend to think of what we hear as being just that: what we hear. It must be sound, coming into our ears, and we just hear it. But it is not that simple. We have developed some highly sophisticated ways of telling differences between sounds that are quite hard to define.

HOW DO YOU TELL THE DIFFERENCE?

Are you familiar with the difference between the sound of various wind instruments, such as an oboe, a clarinet, a trumpet, or a cornet, for example? Then how does the sound of a violin or other stringed instrument differ from that of wind instruments, or from one another? How do you tell all the differences?

And have you ever heard reproduced music in which you could not tell the difference: where you could hear the tune being played, but could not tell what instruments were being played? As you think about all that, you will appreciate some of the nuances that make for clarity in musical reproduction.

Now for speech: when you are conversing with various people in the same room, individual voices are readily recognizable, distinguishable. But how about over the telephone? The same person's voice often sounds quite different over the telephone, doesn't it? And if the line, or the telephone instrument on which you are listening, or he is speaking, is not good, perhaps you could not tell who was speaking, even if you had heard that person's telephone voice before.

So what is it about voice that makes one person's voice recognizable from other voices? Think about that, listen critically, and try to answer this in your own mind. You will find that very instructive.

Reading this and thinking about it, will sensitize you to something that has been going on, all around you, all your life, and that you have taken for granted. If you heard differences, they were out there, not in your head. And of course, they really are out there, but the differences you hear are detected in your head.

So maybe you can detect them, and someone else cannot. Have you taken a car to an automobile mechanic because it was not working right, and he listened to the motor running? He could tell you what was wrong from the sound of it, when you could hear nothing unusual about it. He was obviously hearing something about that sound that you could not hear.

The physical difference was in the real world. But the detection of it by hearing was an acquired capability. That ability takes the form of being able to ignore all the ordinary sounds, while listening for a variety of specific, perhaps tiny, differences that are imperceptible, without training.

In the study of psychoacoustics, a great many experiments have been conducted over the years. Some of these have been directed at finding what differences human hearing can perceive, and what differences are imperceptible.

The result, perhaps, can be illustrated by reference to a well-known commercial for a brand of margarine, which employs the "taste test." A person tastes two samples of bread and something—one butter, one margarine—and is asked to tell which is the butter.

Now, if there were no difference in taste, it would be impossible to tell. But the person usually picks the one that turns out to be margarine, as "tasting more like butter." Assuming the commercial to be genuine (we will not argue if you doubt that), apparently the person tasting can tell at least a slight difference. If you have made that test, you can probably tell the difference quite easily. Some people find it difficult to tell the difference.

The same is true of hearing. But as with taste, a large part of it is training. The finer differences of taste take the sophistication bred by careful use of one's taste buds over many years. How do you think a tea taster, or a wine taster, can do his job? Similarly, appreciation for the finer nuances of sound takes sophistication in listening.

But there is a difference that makes it deceptive: the sophistication we acquire in hearing, is something most of us get, without the conscious effort that we apply to telling differences in taste. We tell the differences between various people's voices, because we want to hear one person's voice to the exclusion of others at the moment. And most often, our hearing develops a rather highly sophisticated ability in this direction, without our realizing it has been happening, ever since birth.

Now, put together all the differences we have been talking about. With the differences between the type of sound to which you may want to listen, the reason why you want to listen to it, and the room or place where you will be doing that listening, you have quite a bit to think about in making a choice of loudspeaker and deciding how to install it for best results. That is what the rest of this book will be about.

Then these differences interact with one another. The kind of sound makes a difference in what you listen for within that sound. Why you want to listen to it can make another difference: are you listening for critical enjoyment, as when you listen to a musical program? Or are you listening to hear what a person is saying, so that it is the message that is important to you?

And where you do this listening makes a difference to how you can pick out whatever it is that you want to listen to or for in the sound of your choice. Different loudspeakers will enable you to get what you want, according to what that is; for different circumstances, it will be different critical listening capabilities.

So let us look next at just what it is your loudspeaker has to do to achieve this variety of capabilities.

2

What Your Loudspeaker Has To Do

As we analyze what it is about sound that enables us to be critical in our listening, the first and most obvious property that causes differences is frequency. Sound is made up of vibrations, traveling through air in a tremendous variety of frequencies, within the range to which human hearing is sensitive.

FREQUENCY

What is a sound vibration? It consists of air particles, moving back and forth, along the direction in which the sound moves, in such a way as to produce minute fluctuations in the air pressure, at our ears, which causes the sensation we hear. This happens because the air through which sound travels is what physicists call elastic, a gas.

This means it is compressible and expandable, and that changes in volume are accompanied by change in pressure. If you squeeze more air into a smaller space, as in a bicycle pump, you increase its pressure. A vacuum pump reverses this process: by making air exit a large space, its pressure is reduced, until eventually you have no air at all, and it is called a vacuum.

Sound transmission does not involve anywhere near that order of pressure change. The changes associated with the sounds we hear are, for the most part, relatively minute—unless you are at a rock concert.

But think about what happens to the air, when (a) some of it moves more than that adjacent to it, and (b) when some of it has a pressure that is either higher or lower than the pressure of the air nearby.

If one little bit of air is moving faster than the particles adjoining it, it will compress the air in front of it, where it is moving to, and expand the air behind it, where it is moving from.

And if one little bit of air is at a pressure higher than the surrounding air, air will move out from that little bit, until the pressure equalizes. If the pressure is lower, air will move in to equalize.

One more ingredient makes sound transmission possible: the fact that, light as it is, air does have weight, or mass. To get something moving, you need force. To stop something moving, you also need force. And in the case of sound waves, differences in pressure are what provide that force. Now we have enough about how air behaves, to see how a sound wave propagates.

Let us illustrate it with a handclap (Fig. 2-1). Bringing your hands together causes an outrush of air. Actually, it is the stopping of that outward movement, abruptly, that makes the sound, but it is easier to think of the outrush that comes before the main impact of the sound. The air moving out increases the pressure of the still air round your hands. This pushes the outward movement further away from your hands, expanding the movement.

But this movement does not happen immediately, because it must build up a little pressure in each layer of air before that layer of air starts the next layer of air moving. As you bring your hands together, the air movement, outward from them, gets faster and faster, until your hands come together, when a fairly rapid movement is suddenly stopped. This is what makes the clap that you hear.

But the air right by your hands keeps moving for a moment, rapdily reducing the pressure there, which pulls the air back in its movement. This reduces the pressure in the next layer of air, pulling back the next layer of air, and so on. You have started an outward moving sound wave.

Now, the kind of sounds we will be mostly interested in, in this book, are more complex than a handclap (although perhaps a handclap is more complex than you realize). Most sounds are characterized by frequency. This means that the back and forth movement of the air particles, and the changes in pressure that accompany them, take place in a rhythmic fashion, so many times a second.

The number of times per second that an air particle in the path of the sound, moves back and forth, or that pressure fluctuates up and down, is called the "frequency" of the sound. The range of

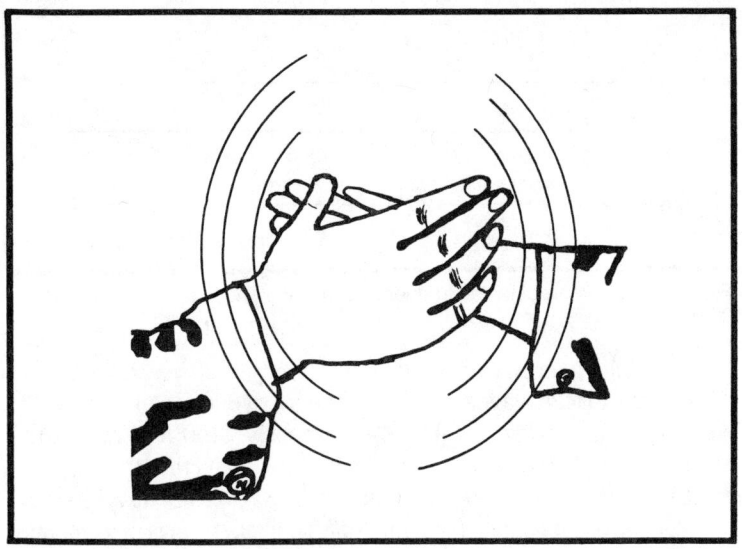

Fig. 2-1. How a hand clap produces sound.

frequencies to which human hearing is sensitive covers a tremendous range.

This range extends from about 20 hertz, or vibrations per second, at the low frequency end, to somewhere between 10,000 and 20,000 hertz, or vibrations per second, at the high frequency end. Individual hearing varies. Some people can hardly hear a vibration of 10,000 hertz. A few can even hear frequencies beyond 20,000 hertz. Most of us, if we are tested on an audiometer, will find that our hearing vanishes at some frequency between these limits.

Most of the sounds to which we listen seem to be well below 10,000 hertz. But these higher frequencies are vital to hearing the kind of details about the sound, that enable us to tell the differences that we discussed in Chapter 1.

WAVELENGTH

Why are we concerned about the frequency of sounds beyond the fact that we want to hear them? Because the differences in frequency make a difference to another property of individual sound waves, or acoustic waves, that is very much connected with how a loudspeaker can reproduce and radiate them—wavelength.

In air, sound waves travel at nearly 1100 feet per second. The exact speed varies a little with atmospheric conditions, and thus the weather. But for all practical purposes, the speed of sound in air is just below 1100 feet per second.

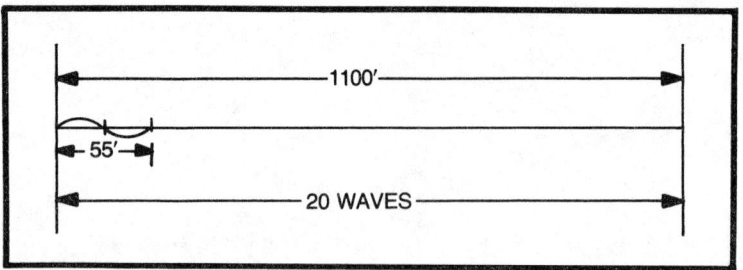

Fig. 2-2. Relationship between frequency and wavelength for a low frequency of 20 hertz.

So a big organ pipe may emit a frequency of 20 hertz, let us say. This means that 20 waves will pass our ears every second (Fig. 2-2). And as the first wave will have gone nearly 1100 feet during the second taken for the 20 waves to pass, this means that the length of each wave must be 1100 feet, divided by 20, or about 55 feet.

On the other hand, a high note on the piano, or on a piccolo, may get up to about 4000 hertz. This means that the same 1100 feet that sound travels every second will, at any instant, contain some 4000 waves, or each wave will be not much more than a quarter of a foot, or 3 inches long (Fig. 2-3).

The low organ tone has a wave that is 55 feet long. The high note has a wave that is 3 inches long. And some of the components in sounds that enable us to tell differences between sounds we hear have wavelengths less than an inch long.

RANGE

That is quite a range of different wavelengths, isn't it? Now think of this. The organ pipe, the piano string, the piccolo, or whatever normally makes the sound in the first place, is designed to make just that one sound. If it is a person's voice instead, that person's vocal cords, throat, mouth, and nose cavity affect the frequency composition of the sound you hear that you identify as being that person's voice.

The point we make here is, each original sound is natural to the source that produces it. A loudspeaker has a much more demanding task: to reproduce all of these sounds so that you are convinced you are listening to the various sources that produced the original sounds. To do this, it must, at the same time, be able to make big waves, 55 feet long, and tiny waves about an inch long, maybe less, as well as all the variations in between, and all equally well.

Further, it must be able to do all this, in the environment into which you put it. First, it must be capable of making big waves, many times bigger than itself, and tiny waves very much smaller than itself, and everything in between. Then it must feed them into the room in such a way that they sound real to you. What difference does that last requirement make?

KIND OF SOUND DISTRIBUTION

In the early days, loudspeaker designers thought that, as long as all those frequencies got out there, the loudspeaker had done its job. Now they know differently. To understand this difference, we need to think about how we listen, in different environmental conditions. In Chapter 1, we talked about live and dead rooms. Let us pursue this.

Perhaps, when you have been moving, you may have talked in a room just after all the furniture and the carpet had been removed, leaving the room very bare. It sounded very different, didn't it? You would not have believed the difference that all that furnishing could make. In your everyday life, you do not seem to encounter this difference. Yet the fact is, you do. You tend to ignore the difference, only because your hearing faculty accepts each room as you find it, as "normal" for that room.

What you may not have realized, is that you always listen differently in different rooms, without knowing it, or thinking about it. The dead room makes for easier listening, because you do not have to concentrate so intensely, to hear what someone says, or to sort out the music. At the same time, in the live room, you make do by training your hearing to ignore as much as possible except the sound you want to hear.

Most of us have become so good at this that we do not realize how much our hearing can and does ignore, until we think about it. Mother's sometimes accuse their children of not hearing them, an accusation that seems impossible unless the children's hearing is defective. The mother accuses them of hearing "only what you want

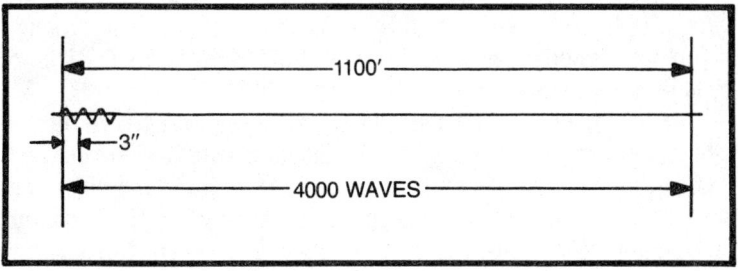

Fig. 2-3. Relationship between frequency and wavelength for a higher frequency of 4000 hertz.

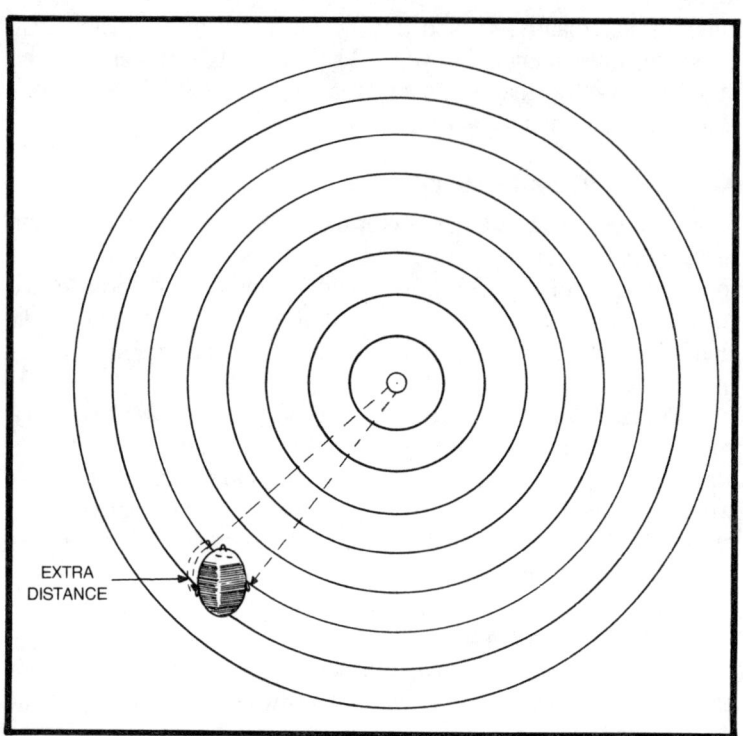

Fig. 2-4. How our binaural hearing faculty gets the information to tell us in which direction the sound is coming from.

to hear." Unlikely as it may seem, we all do that, far more than we realize without even thinking about it. This fact is very important to the purpose of this book.

LISTENING DIRECTIVELY

Our binaural hearing—the fact that we listen with two ears—is an important part of the mechanism by which we can listen directively (Fig. 2-4). And we do it without even thinking about it, generally, because we learned to do it while we were very young, before we learned to talk. So, even if we think about it, we find it hard to believe that we are doing it.

Have you ever taken a small tape recorder with you to record what happened at a meeting you attended? It sits beside you, and "listens" to what you listen to (Fig. 2-5). During the meeting, you were able to hear everyone, except perhaps a few who muffled their voices. But when you replay the tape, the only ones you can hear, or understand, are those who were close to where the tape recorder was.

You can turn up the volume, but even then it is far more difficult to make out what was said than it was for you sitting there during the meeting, because the recorder amplifies all kinds of other sounds that you never noticed were there, at least most of them. But they were there, or the recorder would not have picked them up.

The tape recorder just does not have your ability to pick up only what you want to hear, while rejecting all those unnecessary sounds. Your hearing faculty has a sort of built-in programmer that processes everything your ears pick up, which is the same as that tape recorder picked up, to let you listen to just what you want to "hear," while rejecting everything else, what you do not want to hear.

Your sophisticated hearing faculty—not just your ears—has become so adept at processing the sounds your ears pick up, that you do not even think you hear all those extra sounds that the tape recorder picked up. Why could not the tape recorder sort it out for you, the way your hearing does?

And more importantly, for the purposes of this book, why can't you sort it out, as easily as you could when you were there in person, as you listen to the tape recorder play it back? Because you have lost something important in the process.

Your ears sample the sounds that pass your head at two locations: one on either side of your head, where your ears are. The tape record just took one sample, and put it on its one sound track. Your ears bring your brain's hearing faculty two samples to work with. The tape recorder has only one. And although you listen to it with

Fig. 2-5. A listening enigma: does the tape recorder hear the same as you do?

two ears, they both hear the one sound track the tape recorder reproduces, not the two samplings of sound that your ears had, when you were there.

If you are in a room where a lot of people are talking at once, you listen to the ones you are interested in, do you not? You get involved in one conversation and ignore all the others. Then perhaps you hear your name from someone asking, "What do you think?"

You were not listening, so you do not know what they are asking, do you? And you realize that, had you been paying attention, you could answer the question, but you do not know what it is, because you were not really listening.

Forgetting, for the moment, the embarrassment that such a situation may cause you, think about why something like that can happen. You were awake, perhaps listening quite intently to something else. Your ears must have picked up the information you should have heard. But your brain conveniently ignored it, in that wonderfully selective hearing interpretive faculty of yours.

The fact is, every one of us is doing this, every moment of our lives, without realizing it. Now, relative to our problem of picking an appropriate loudspeaker, what changes, from one room environment to another, is the degree to which our hearing faculty has to do this kind of selective listening thing for us, to enable us to hear what we want to.

In a dead room, our hearing can relax and does not have to be nearly so selective. In a live room, our hearing works overtime, singling out what we want to listen to.

LOUDSPEAKER DIRECTIVITY

The loudspeaker system is at the other end of this same situation. Your ears are receivers, while the loudspeakers are transmitters. They both have to deal, in different ways, with the same situation. In the dead room, very little sound is reflected from walls, floor, ceiling, and other surfaces. So you must hear sound directly from the loudspeaker, wherever it is located.

In the live room, sound gets reflected a lot, so in this situation you want a loudspeaker system that allows in some way for this fact. The reflections should seem normal for that room, and interfere as little as possible with your ability to hear. We will get into that more in the next chapter.

We have talked about reflection and absorption of sound waves. What causes that, or makes the difference? To understand this, in its simplest terms, you must realize that sound waves are a form of energy, although quite small, generally speaking. When air is compressed, compared to the surrounding air, that represents a small store of energy, what physicists call "potential energy." When air is in motion, due to the passage of a sound wave, that represent a small store of another form of energy—what physicists call "kinetic energy."

While sound waves are traveling outward across air, the energy keeps on going with very little loss. It may expand to fill a larger volume, but it does not get lost. But when a sound wave encounters a wall, or other barrier made of something other than air, something different must happen there. And it is what happens there that determines whether reflection or absorption occurs.

Actually, this depends on the material of which the wall, ceiling, floor, or whatever other barrier it encounters is made of. The sound wave is never totally reflected, nor totally absorbed. But some substances can come pretty close to doing one or the other, while some substances do quite a bit of each.

If the substance is quite rigid, like a concrete floor, it is not about to move in response to a sound wave. So the arriving air particles, due to the sound wave, get stopped in their tracks. This doubles up the increase in sound pressure, as compared with what happened when the wave was still moving in air, and the increase starts a wave back into the air. This is called reflection.

The total effect of reflection, at such a hard surface, is very like what happens to light, when it strikes a mirror. A sound wave starts back into the air, at an angle "bent back" just as light waves are in a mirror.

Now what causes absorption? If the particles of the substance that the wave strikes, move in response to the changes in pressure, then they will absorb some energy from the sound wave. The pressure will not build up so much, so there will be correspondingly less reflection.

That is what happens to sound waves in different kinds of room, which you will have observed by now even if you were unaware of it until we called it to your attention. But what has that to do with your loudspeaker's job?

For now, the point to realize is that loudspeakers can be directive in the way they distribute sound, as well as being able to send out all the various frequencies required. The desirable degree of directivity, and the way it is used will vary according to the room's characteristics.

MONO, STEREO, QUAD

This directability quality of loudspeakers must also be correlated with the kind of system you want to use them in, whether mono, stereo, or quadrasonic. Back in the days when designers of systems first realized that frequency response was not everything, and that directional perception plays an important part in everyday listening, they sought to provide such directional perception by means of stereo.

Although we listen with only two ears, the ability to reproduce, or recreate, a sound pattern around our heads that reasonably represents what we might have heard at the original performance (whatever that was) is not as simple as just using two loudspeakers instead of one.

BINAURAL

To understand why this is, we should distinguish between binaural listening and stereophonic reproduction. Some of the earliest demonstrations of stereophonic effect used binaural listening by means of headphones. This was first done back in the Nineteenth century, believe it or not, quite successfully.

If a dummy head, shaped like a human head, has a pair of microphones installed where the ears should be, and the sound each of them picks up is piped to the corresponding earpiece of a pair of headphones, the sense of realism is fantastic (Fig. 2-6).

The only thing unnatural about it, is that you do not normally listen to everyday happenings with a pair of headphones on. A particularly unreal feature is that every time you move your head the whole room (in which the original sound was picked up) seems to move with your head because the headphones are the source of everything you hear. Turning your head does not change your aural perspective as it does when your ears are free.

Think about that. In your normal listening experinece, turning your head does not affect the sound waves that keep buzzing round it; however, moving your head will change the points at which your ears sample all these sound waves. But you have come to expect this change. It is what normally happens when you move your head.

Now, you put on a pair of headphones, to listen to a binaural recording, or program of some sort. Now, when you move your head, the headphones move with it. The sampling of sound, conveyed artificially to your two ears, does not change with your head movement as it always does when you are not wearing headphones. So the result is unnatural.

To overcome this defect, designers wanted to do the same thing with loudspeakers. But that is not so easy. As soon as sound is let loose into the air, the characteristics of the room get into the

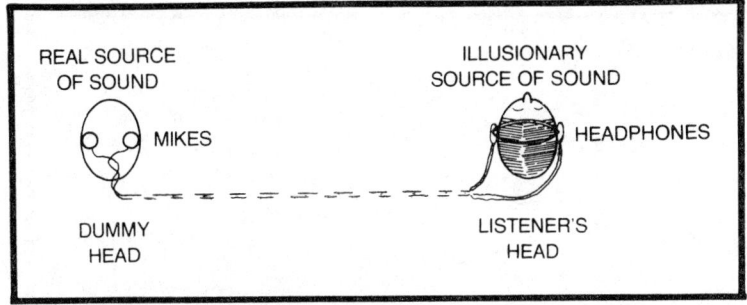

Fig. 2-6. Basis for binaural listening experiment. This was first tried, very successfully, at the Paris Exposition of 1889.

picture, in the ways we have been discussing. Headphones did keep that out but had their own unnatural effect.

EARLY STEREO

The original concept of stereophonic sound, to give stereo its full name, sought to connect the room in which you listen with the room in which the original performance took place by using a number of channels to simulate removal of what was really two walls, but by this means became a single imaginary wall. Sound was thus let through from the first room to the second.

To achieve this, each channel, with its loudspeaker, reproduces what is received by a corresponding microphone. The effect, each channel carrying its own piece of the sound wave reaching that wall, is as if the two walls coincide, and sound passes one way through it (Fig. 2-7).

The word "stereophonic" means solid sound. It implies that you recreate a solid field of sound that is as nearly as possible a replica of the original sound field, extended into the room where you listen. This basic idea got simplified down, in the case of movie theatres, to three channels, and for home use, to two (Fig. 2-8).

Technology is always advancing. When stereo first came in for home use, the quality of reproduction was nowhere near as good as it is today. The best systems then may have been better than poor systems today, but on the average, quality was much poorer then.

This inferiority of quality meant that people who listened, to compare mono and stereo, could not so readily tell the difference between the two. In many listening situations, stereo reproduction sounded indistinguishable from putting the same two loudspeakers on mono. It does take good quality stereo program, played on a good quality system, for two loudspeakers reproducing stereo to noticeably improve on mono reproduction.

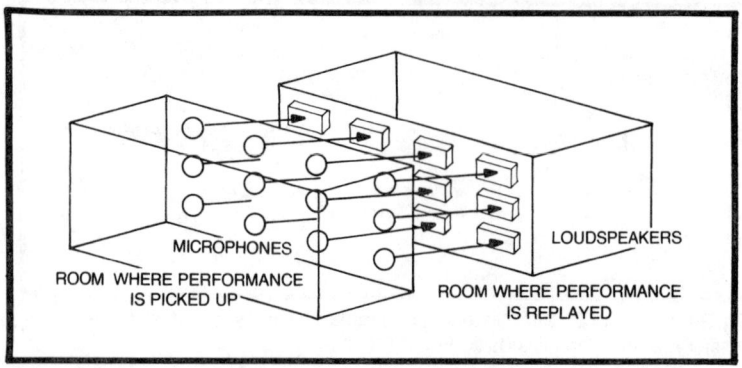

Fig. 2-7. The basic concept of the first experiments in stereophonic sound.

Because of this, the early demonstration material tried to overcome the deficiency by overemphasizing the stereo effect during recording. Different instruments, or different vocalists, would be entirely on one channel or the other, so that it was like listening to two different, but related programs, one reproduced over each loudspeaker. Then they would have one instrument do a piece, then the other, a procedure that caused people in the business to call it the Ping-Pong effect.

While such recordings did dramatically demonstrate that stereo enabled sounds to be separated, it did not really exploit the true advantage that stereo posseses over mono in everyday use: how it can help us hear with better realism, everyday sounds or programs, which are not naturally "Ping-Ponged" all the time.

But before we move on, note a particular defect that is inherent in the arrangement of Fig. 2-7 as compared with a true live performance. We have already mentioned it, but only in passing, and you may have missed it. The sound can go only one way, in the theoretical reconstruction, from the studio to the listening room. There is no way by which the listening room can really become part of the original scene.

Our hearing faculties naturally distinguish different sounds by isolating different groups of frequencies associated with various sources of sound, along with the different directions from which they arrive at our heads, to deliver complex impulses to each ear. For this to work, there has to be a separateness about the various sounds.

Early stereo advocates believed that this separation should enable the listener to pinpoint the direction of any particular source with high accuracy, or that realism of reproduction depends on this capability. While these properties may indeed be related, the fact is that we do not normally spend our time trying to pinpoint everything we hear. But we do separate individual sounds, whether they be voices, musical instruments, or whatever.

STEREO DEFICIENCIES

Two things were recognized as being deficient in the early home stereo systems. First, if the left and right loudspeakers were widely separated, there seemed to be a hole in the middle. And second, if both of them were in front of where you were listening, there seemed to be an emptiness behind.

Coupled with the emptiness behind effect, was the lack of what is now called *ambience,* due to the reverberation that characterizes what you hear, particularly in a large auditorium. When you are really in the auditorium, this room reverberation comes from all around you. But when the same program is reproduced in a much smaller, less reverberant room, and reproduced entirely over two loudspeakers located in front of the listener, all of the sound, includ-

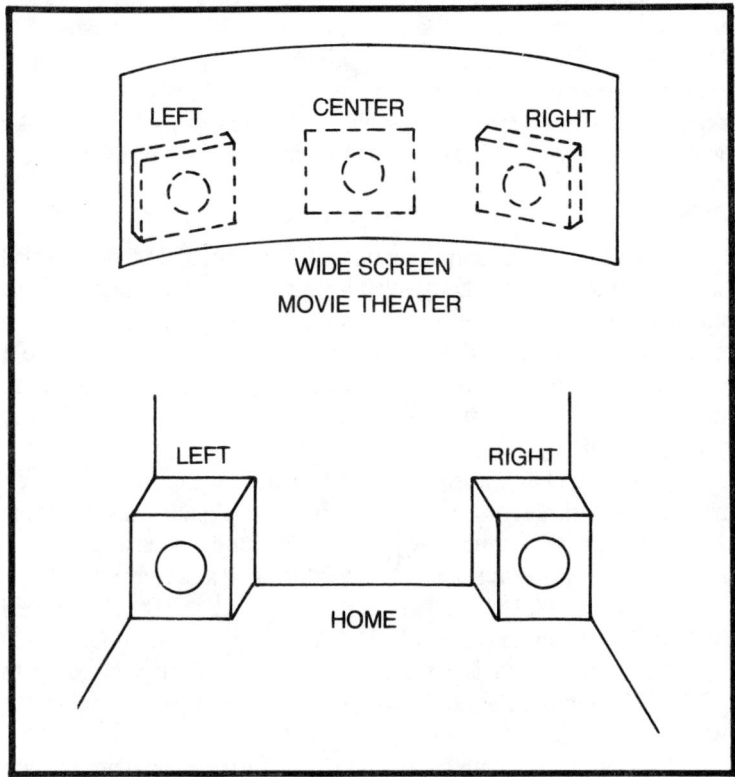

Fig. 2-8. Early practical stereophonic sound systems, as used for the movie theater and the home.

ing recorded reverberation, comes from in front, leaving an emptiness behind.

Quadraphonic was introduced, mainly to eliminate this deficiency. How and where it is most effective, we will have more to say about later. The point to realize here is that the system for which you want the loudspeakers can make a difference to the choice you make, whether for mono, stereo, or quadraphonic.

EFFICIENCY

One more factor needs to be considered to get all the cards on the table before we really get down to details. This is loudspeaker efficiency. We often hear people talk about how many watts their system has. Someone who has a 100-watt system seems to think he has twice as much power as a person whose system rates only 50 watts.

In some instances this may be true with regard to the effect of the system on his electric light bill, although that is not even a reliable indicator. But relative to the system's performance it means very little without taking other factors into account. But even assuming that was the whole story, doubling the power, everything else being equal, raises the sound level by only 3 dB, a difference for which you would have to listen quite closely to even detect a change.

POWER

If the ratings were in acoustic watts, rather than the electrical watts used to produce them, such a rating might have a little more meaning. But to put even acoustic watts in perspective, we need to recognize that the sound level, which should be measured where it enters our ears because that is what we hear, can occupy a very wide range of level indeed between a sound that we can just hear and one that is unbearably loud.

The level of sound that begins to cause pain, and this varies from person to person, has a level that averages a trillion (that is 1,000,000,000,000) times the power in a sound that is just audible. Acousticians use the average just-audible sound as a reference level.

A comfortable level might have an energy 1,000,000 times just-audible or threshold level. That is 60 dB. If you double it to 2,000,000 times, that is 63 dB, only 3 dB more. If you go up to 10,000,000 times threshold, that is 70 dB, or 100,000,000 times would be 80 dB.

Now, suppose that the efficiency of the loudspeakers is such that 1 watt electrical power produces a 60 dB level. It will take 2 watts to produce 63 dB, 10 watts to produce 70 dB and 100 watts to produce 80 dB. As far as listening is concerned, the change in level going from 1 watt to 10 watts is about the same as going from 10 watts to 100 watts.

If you are new to what *dB* really means, this may still be a little confusing. Decibels, for which dB is an abbreviation, are a way of measuring relative level. They must always be relative. Thus one sound can be 3 dB, 10 dB, or whatever louder than some other sound. In general use, to say a sound is 30 dB or any other figure does not mean anything, although we have heard such an expression used, and we will explain that in a minute.

As we have just shown, a 1 dB change in level is audible only by very careful listening on a steady tone. A 1 dB change in program level is definitely not an audible change. Thus, professional faders that serve like volume controls actually change level in 1 dB steps, not completely gradually at all. Some of them even use 2 dB steps. And you'd have to be a very well trained critic to hear even those steps. Such changes appear completely graudal, as they build up to something that is audible, like 10 dB.

But now we were talking about decibel levels. This is a completely arbitrary term but is usually taken to mean above the threshold of hearing or a sound just barely audible. We should add that there are very few places in the world today where you could hear such a sound, because there are so many sounds that would drown it out.

When we refer to the threshold of hearing, we mean against a background of complete silence, something very rare in today's world, anywhere. Against a background of complete silence, normal human breathing (not when you are out of breath, or suffering from a fever) is apt to sound like a bull in full chase.

Now, with that as our reference level, a comfortable listening level might be about 70 dB above that. We suggested, in suitable circumstances, that might take 10 watts. Then 20 watts would raise the level only to 73 dB, and the change would be audible, only with very careful listening. To get a level comparable with the sound of a jet aircraft as heard from the waiting room when the jet is on the runway, you would need about a million watts of power or more.

You see why doubling the power does not mean too much, without taking other factors into account. But are we going to talk about electrical or acoustical watts?

To consider total acoustical power, you need to consider the volume in which that power is developed, or generated, and the absorption factors of the room or walls surrounding the space or volume. The more absorption there is, the more acoustical power you will need to achieve a corresponding loudness.

But to consider electrical power needed, we must also know loudspeaker efficiency. As far as the loudspaaker you will eventually pick is concerned, it can have a wide range of efficiency values. What do you think you would rate as a high efficiency loudspeaker: 90%? If you think that, we have news for you.

The highest efficiency loudspeaker ever made ran at an efficiency slightly over 50%. In everyday terms, that was superefficient. What today is called a high efficiency unit will run somewhere between 10% and 20%. In fact a unit that has an efficiency of 20% is very high compared with most others.

The average commercial loudspeaker has an efficiency between 2% and 5%, and many of the low efficiency units are far below 1% efficient. So when you talk about 50 watts of power delivered into low efficiency units, the sound energy fed into the room is only a small fraction of an acoustical watt.

With high efficiency units, between 2 and 10 electrical watts peak output is plenty for the average sized living room. When we say plenty we mean enough to lift you out of your seat on peaks. But put the same electrical output into low efficiency units and the same power will seem very inadequate.

So efficiency is important, particularly when you come to talk about the size of amplifiers you need. Efficiency is also important because the different types of loudspeaker radiate their sound in different ways.

All this adds up to the need to know exactly what you are doing, before you decide on the best way to go about it. We want to stress that you should read the rest of this book carefully before you jump into buying and installing units for whatever purpose you have in mind. The following is a comparison table of electrical versus acoustical watts.

Efficiency(%)	**Electrical Watts**	**Acoustical Watts**
20.0	5	1.0
10.0	10	1.0
2.0	50	1.0
1.0	100	1.0
0.1	1000	1.0
0.1	50	0.05
0.1	10	0.01

3

Fitting a Loudspeaker To Its Job

So what is the job you want your loudspeaker to do? What size and type of room, what kind of program, and so forth are involved? By now, you should have the questions well in mind. You need to think in terms of how your hearing normally functions in whatever the situation is.

As the previous chapters have shown, you spend all the time you listen separating what you want to hear from the context in which your hear it within the limitations of your hearing capability. Only if the context produces so much confusion that you find it impossible to sort the sound out, do you conclude that the sound is bad in that room or auditorium.

And even in a bad room, careful listening can overcome your problems to an amazing degree. Many of us have been to one of those old theaters designed for silent movies that were never properly converted for talkies. When we first entered the sound seemed impossible. We could not distinguish a word that was said because of reverberant sound bouncing around. But, having paid admission, we sat there a while before deciding to leave and, after a while, found we could decipher most of what was said, missing a word only now and then.

THE POWER OF SELECTIVE LISTENING

Did the room actually improve while we were sitting there? No, our hearing became accustomed to it, meaning that we acquired the ability to ignore more of the unwanted sound, thus extracting more of what we wanted to hear. Relevant to the purpose of this book, the point is that listening will be much easier, more comfortable, if we do not have to strain, consciously or subconsciously. But it is equally important to realize that we do make such allowances, usually to a lesser degree, all the time without even realizing it.

How does our hearing do that? It is possible because we acquire the capability to separate original sound from reflected, or reverberant sound. If you study the differences, you will discover that this is mainly due to the fact that we can tell the direction from which sound arrives at our heads.

So we are able to concentrate our attention in the direction from which the sound we want to hear comes. We do this whether this is where the loudspeaker is, or whether it is where the person is whose voice we want to hear above the general hubbub of conversation going on.

But while directional listening helps, it is not all we use, as our hearing becomes more proficient. And if you are one of those people who have good hearing in only one ear, you cannot use that faculty at all because directional hearing depends on having our brain compare the sound waves picked up by each ear very carefully.

The other part of our ability to separate sounds from different sources is related to that binaurally based capability, a critical analysis of the composite sound itself. If you are listening to a particular person's voice, you note the idiosyncrasies of that person's voice. If you are listening to music, you note the characteristic tone qualities of individual musical instruments.

When you listen to reproduced sound, you depend on loudspeakers to recreate a sound field around your head from which you can get information similar to what your hearing would have obtained from the original sound field.

Earlier theories about sound reproduction suggested that perfect reproduction would recreate a sound field, around each listener's head identical with the original. Further studies showed that this is, at the present state of the art, impossible. Why?

ACHIEVING THE OBJECTIVE

Because you listen to sound in a room of some sort. What we have been showing you, is that everything you hear in any room, whether it is original sound or reproduced sound, is affected by being heard in that particular room. Your hearing adapts to the characteristics of that room, whether you ever realized it before or not.

We have given some examples that show how your hearing is affected by the room you are in. This same variation was particularly troublesome to the developers and designers of loudspeakers and microphones. They could not get away from the effect of the room they tested them in. And if they tried testing them outdoors, the sound of birds, passing aircraft, and local traffic posed problems.

So audio and acoustic engineers developed a device called an anechoic room or chamber, that is, a room completely without echo, or sound reflection. It is built with long wedges of very highly absorptive material, such as fiberglass foam (which is really about 99% air, but contained in a cellular structure that makes it different from free air).

These wedges may be 4 to 6 inches square at the base, where they fasten to the walls, the ceiling, and even the floor (what you walk in on and put things on is actually a strong mesh that provides very little obstruction to sound waves). They are 2 or 3 feet long and go to a point where they reach the room space.

A sound wave approaching the wall, first comes in contact with these thin points, which absorb parts of the wave. As the wave progresses into the space between the wedges, more of it gets absorbed, by the thickening mass of fiberglass foam. When the sound wave would have reached the actual wall of the room, it has been totally absorbed by the fiberglass foam. This is the nearest thing to total absorption that there is.

Using a microphone and loudspeaker in such a room, enables the engineers or experimenters to test either one, and find out exactly what it does, without any room effects. Would that be a good place to listen to your favorite program recordings? More of that in a moment.

This means that you cannot reproduce a sound field identical to some original sound field that was produced in a different room, because you cannot get rid of the effects of the room in which you listen to it. If you refer to Fig. 2-7 again, really removing the intervening walls would allow sound reflections to pass both ways. All that the most elaborate stereo system can accomplish is the effective passage of sound one way only.

What this means is that you must accept the room in which you listen, as it is, unless it is too difficult to listen in at all, in which case you may have to do something about that. But even after such treatment, you will have to accept the room as it then is with its treatment. You do not, cannot possibly, try to make the room in which you listen identical to each room in which individual programs to which you want to listen were originally produced.

In some recording studios, they do in fact approach this capability, however. Studios have been built with movable, or swingable

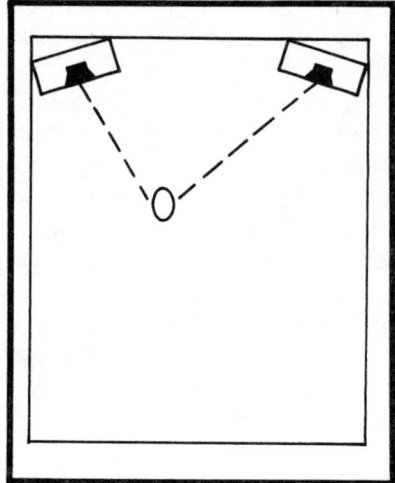

Fig. 3-1. An arrangement that is usually good for a dead room, one that is furnished so as to be highly absorbent of sound.

panels. In one position, the walls are almost totally reflective. In other positions, parts of the walls, or perhaps most of them, become much more absorptive. Using this kind of construction does enable one actual studio to be used in different ways, so that its accoustic properties can be totally different.

But this is not a practical thing to do to listening rooms, many of which double as living rooms, or playrooms. In this case, we most often have to settle for what we have, and try to tailor the loudspeaker system to suit the room we have.

ROOM QUALITY

To illustrate what this difference means, consider first a stereo installation in the two extremes of room type, but of the same size: one a highly absorbent, or dead room and the other a highly reflective, or live room.

In the dead room, the ideal arrangement is for the loudspeakers to face into the room, in such a way that anyone listening can hear each loudspeaker directly (Fig. 3-1).

In a live room, one arrangement that is often successful uses reflection deliberately (Fig. 3-2). The effect of any highly reflective room, is to make it sound bigger than it really is, because sounds often travel farther from source to listener than in a dead room of the same size.

If you have ever visited an anechoic room, in which all surfaces are made as absorbent as possible, like a superdead room, you might think that the complete lack of reflections would make it seem like outdoors, where there are no walls to reflect. But it does not.

Instead, the sort of eerie silence gives you a sense of aural claustrophobia, even if you are not normally claustrophobic.

Perhaps we should explain why an anechoic room is so different from open air, since both would appear to have no reflections. Before you even start listening to something, or trying to talk against it, the first thing you notice is the silence. Outdoors, there are invariably some sounds. If it is quiet and peaceful, say out in a redwood forest, there will be the birds, or the rustle of a leaf, as some animal moves it. It is never completely silent. In more civilized outdoor situations, there will be a great many more noises.

But in the anechoic room, it is silent, except for sounds that you make. And even they seem different. Outdoors, even in the quietest place you have ever been, you have the ground beneath your feet. Its reflective characteristic will vary, according to whether it is rock, grass covered, or perhaps covered with pine needles or leaves.

When you speak, or even breathe, the sound comes back to you from the ground whether you have ever noticed it or not. Often there is too much other sound for you to be able to notice it. But in a really quiet background you will. In the anechoic room, there is no reflection even from below you. It is as if every sound you make, even breathing, is sucked up by the room, making a most unnatural experience.

But back to the practical situations we face. We have a deal with rooms that are only comparatively live or dead.

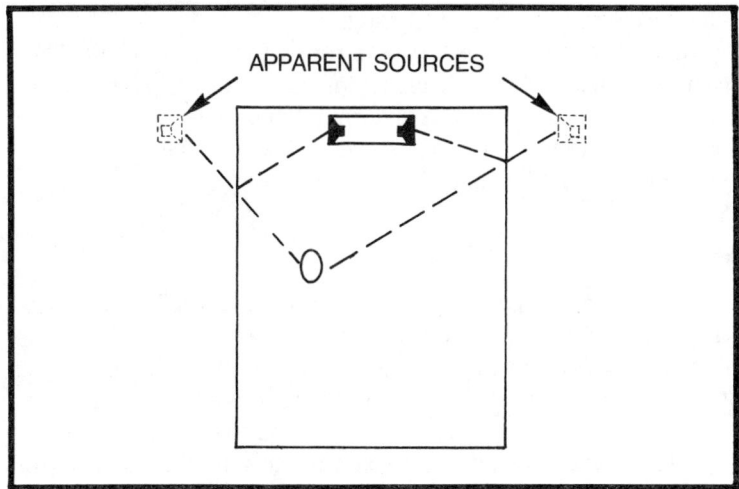

Fig. 3-2. One arrangement that often serves a live room well, one that has most of its surfaces highly reflective of sound.

So, in the live room, by using a directive rather than a diffusive type of loudspeaker, and directing its sound toward the wall, what the listener hears gives the impression of sound sources farther apart to match the apparent room size.

Note the difference in loudspeaker requirement. In the dead room, the loudspeaker needs to diffuse, and to have a frequency response that is widely diffused, so a listener anywhere in the room will hear a good response direct from the loudspeaker. In the live room, it is often better if the loudspeaker confines radiation to an angle that prevents sound from reaching the listener by a direct path, or at least makes the indirect path the strongest component the listener hears.

Understanding the reason for this difference in treatment is important. In the dead room, although not to the extent that would be experienced in an anechoic room, sounds do not reflect much. So it is necessary for the listener to hear the sounds directly from the loudspeakers. In most rooms, people may occupy different positions (not what at one time was regarded as the ideal stereo position). They should be able to hear good quality sound, wherever they happen to be.

This means not only that loudspeakers for this use should diffuse the sound so it is heard in all directions in every part of the room, but that it should serve all the frequencies it reproduces uniformly in that respect. A loudspeaker that distributes the lower frequencies fairly uniformly, which most do, but becomes directional at the higher frequencies, will give quality of reproduction that differs according to where you listen.

In the reflective, or live type of room, the requirements differ again. Now the objective should be for the various frqeuency components to be heard, either as uniformly as possible after they have bounced off the various surfaces by which they reach the listener, or in a manner as natural as possible to that room.

ROOM SIZE

So much for quality of the room. How about size? Now we also have to think about what you want to hear because the size of the room affects realism with many kinds of sound. For example, if you are listening to a concert, you want the impression of an auditorium even if you are listening in a much smaller room.

On the other hand, if you are listening to something that has, or should have, a sense of intimacy, such as a play or a conversation, the spaciousness of an auditorium destroys the desired illusion. You want an environment that would lead you to think that only you are listening.

What can give you that illusion? In a real conversation, you are relatively close to the person to whom you are listening, so you hear his voice directly. Also, unless it is an argument causing him to shout, he talks in a conversational tone. Now, on stage, play actors have to do something different.

In theaters that do not have sound reinforcement, they must project their voices so that the audience can hear them, while maintaining a tone of voice that remains as conversational as possible. Actors and actresses who can do this successfully have had considerable training as a rule.

Most modern theaters use sound reinforcement. And to get good realism with a play, the microphones must pick up the sound so it sounds as much like conversation as possible. And the loudspeakers must project it toward the audience so they all hear it directly without any more added reverberation than can be avoided.

To achieve this, reproduction in the theater must be of the highest possible fidelity, and with the least possible reverberation from various wall and ceiling surfaces. In this way, the quality of voice that the microphones pick up, is beamed as intimately as possible, to the ears of the audience. But that is not the only effect that room size, or apparent room size, can have.

Size of room also affects how the loudspeaker handles different frequencies in the audible range, particularly at the lower frequen-

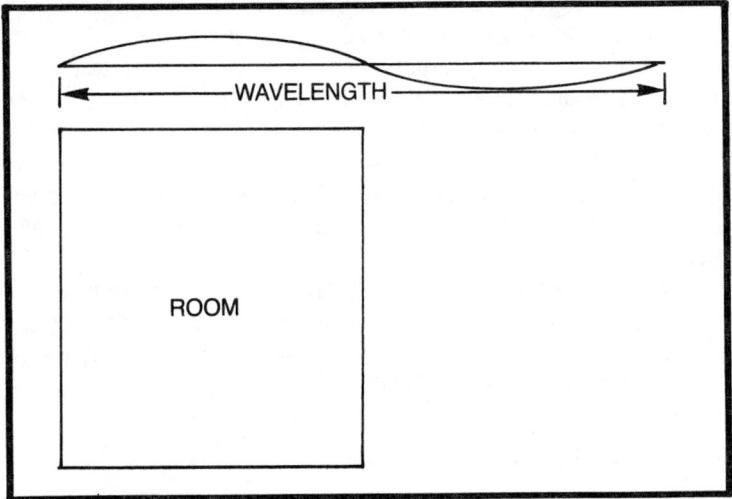

Fig. 3-3. The long wavelengths associated with low frequencies mean that the average room can house only part of a wave at such frequencies.

37

cies. What we were saying just now, about reflective or absorptive properties, is usually more important to the higher frequencies, above 500 or 1000 hertz. Size of room has more effect on how we hear the lower frequencies, say below 250 hertz.

At really low frequencies, a smaller room can house only part of a wavelength of the sound you hear (Fig. 3-3). Obviously, under that circumstance, the sound wave cannot travel in the usual sense. What you get is just a pressure fluctuation within the room. At the same time, the higher frequencies still travel across even the smallest room and get reflected or absorbed as the case may be.

Where this change of effect takes place depends on the size of the loudspeaker box and the size of the room. At any frequency lower than one for which the average room dimension is a half-wavelength, all the loudspeaker does is to fill the room with a varying pressure, at the frequency in question.

Thus, if a room is 8 feet high, by 16 feet by 20 feet, its average dimension is about 13½ feet. If 13½ is half a wavelength, a full wavelength is 27 feet. This corresponds to a frequency of about 40 hertz. Any frequency up to 40 hertz, in this room, is produced in this manner. If the room is smaller, the frequency will extend correspondingly higher.

The next point at which a change in the way sound is radiated occurs, is related to the size of the loudspeaker box and the way it is built, which affects how the sound comes out of it, which we discuss fully in later chapters.

So different things happen at different frequencies, as we need to be aware of this difference, if the sound is to "go together," so it sounds like one sound, that belongs together.

Going back to the low frequencies again, the fact that for a certain range the speaker merely fills the room with sound pressure, rather than actually propagating it in the way that higher frequencies are handled, makes things easier all around. The system represented and described in Fig. 4-14 will describe an excellent example of how this property is utilized to achieve reproduction in a quite different way.

But whether an installation applies such a unique strategy as that one, or whether it is more conventional in its design, the fact that the low frequencies fill the room, rather than have any kind of source identification, is one that you can use in less sophisticated ways.

THE CAR AS A ROOM

Take an automobile, which is about the smallest room size you will ever want to listen in. It may be 5 or 6 feet wide, from 6 to 8 feet long, and 3 or 4 feet high from floor to ceiling. Thus it has a total volume, somewhere between 80 and 200 cubic feet. Taking the average dimension as about 5 feet, this is half a wavelength at a frequency of about 110 hertz, or a wavelength at a frequency of about 220 hertz.

This means that, below a frequency in this region, any sounds heard inside the car will be due to pressure fluctuations, not due to any normal sound wave travel. Now, if you open the window, you will allow these pressure fluctuations to escape, changing what you hear in the low-frequency range. And the sound heard outside will be quite different from the sound heard inside the car.

If you are concerned with getting consistently good response from a car loudspeaker installation, you should settle for keeping the windows closed, or at some particular setting. But that may not always be practical. And many teenagers want to let the sound out, so others can hear it, to advertise their presence, their equipment, or whatever.

To get a better perception of the problem of getting good reproduction in a car, when you realize that some of the larger loudspeakers have cabinets not that much larger than a car, you could almost say that listening to sound inside a car resembles being seated inside your loudspeaker (Fig. 3-4).

The difference is that with a conventional loudspeaker, however large it may be, nobody sits inside it, and you want to hear the sound outside, so such a loudspeaker is designed to feed the right kind of sound to the outside world. But in an automobile you are sitting inside, so you are concerned with what it sounds like there rather than outside.

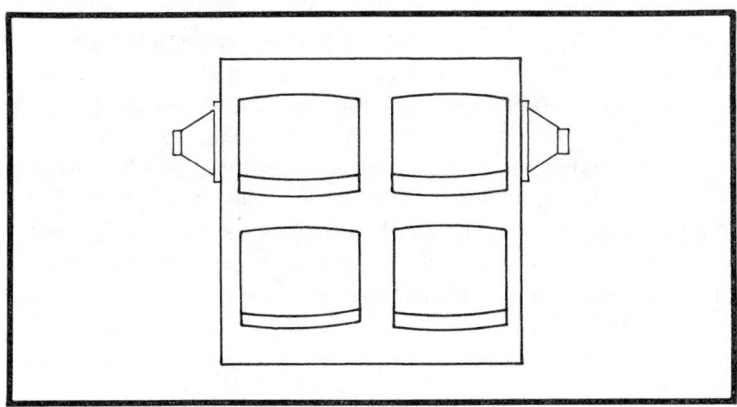

Fig. 3-4. One concept of providing reproduction in a car. Your listeners are seated inside the speaker box.

Most modern automobiles, in the interests of giving their occupants a quiet ride, are furnished with sound absorbent material. This puts them in the dead-room category. Loudspeakers should be placed where they can radiate the higher frequencies so that all occupants of the car can hear them directly.

However, that ideal would create practical difficulties by putting the loudspeakers where they would interfere with necessary vision of the outside world, hampering driving. But a popular place for mounting them overcomes this problem in a simple way (Fig. 3-5). Whether the units are mounted in the front or back deck (or both), the sound will be reflected from the windshield or the back window, which is slanted.

Notice that, as in the method used for a live room (Fig. 3-2), the reflection creates an aural illusion of the room being bigger than it really is, which is an advantage when the actual size is as small as an automobile.

Another place to mount loudspeakers in a car, is in the paneling of the vehicle, so the back of the unit is between the inside and outside shells of the vehicle. If this involves placing it below the window level, it means that sound will reach the listeners' ears by a very indirect path, which is not good but may be adequate. If it can be mounted in a space above window level, that is better, but it depends on such a space being available in the car in question.

Although it may not seem so obvious, doing a good job in a car is also affected by the same factors that must be considered in larger rooms: program choice and mode of reproduction—mono, stereo, or quad. In fact the possibilities for getting unique effects in a car installation can be quite intriguing.

The question reduces to whether you wish to create the illusion of a wide, voluminous, concealed source of sound, or whether you want a more conventional pinpointing effect. The average car is almost a built-to-order listening booth in which you can install a variety of experimental systems with quite interesting effects.

Usually, mounting the loudspeaker units in the paneling of the vehicle avoids low-frequency problems, because the space between the shells communicates with a fairly large volume, allowing the front of the unit to fill the car body with low-frequency sound pressure that is adequate to give the impression of good bass response.

What we have been seeing here is that the performance of a loudspeaker must suit its environment, and must do that at all audible frequencies. It will often be found that the way it achieves this varies for different situations. But as well as a performance match, it must also fit physically into its environment.

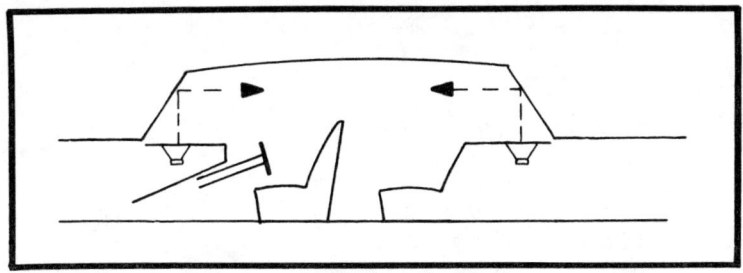

Fig. 3-5. A popular method of mounting automobile loudspeakers uses the windshield and rear window as sound reflecting surfaces to advantage.

PHYSICAL FIT

For example, in a large room, large cabinets containing loudspeakers of one or other type, which we will describe more fully in the next chapter, can easily be housed as part of the room furnishings. But in a smaller room, a loudspeaker cannot sensibly be larger than one of the so-called bookshelf types, simply because the room would not allow a place to put anything bigger.

And in an automobile, space is even more restricted, so the loudspeaker must fit into some space where room can be found for it. If you have got the picture of the various ways in which a loudspeaker must fit into the environment in which it must do its job, we can move on to the next chapter, which shows how different types of loudspeakers go about doing their job.

4

How Each Kind Does Its Job

Before loudspeakers of any kind were invented, the only kind of electrical reproduction available was from headphones. In those days, someone discovered that putting headphones in a pudding basin amplified the sound, so that several people could hear from one pair of headphones at least something, if not all that well (Fig. 4-1).

Then, of course, electronic amplification and loudspeakers were invented to reproduce sound more effectively. We have come a long way since then.

MOVING IRON

The first loudspeakers were of the moving-iron type, in which the electrical currents corresponding to sound waves produced a varying magnetic force that moved an iron reed, to which was attached a diaphragm usually made of paper or parchment.

This type had very restricted performance, being little better than an improvement on the headphones-in-the-basin sound. The first real step toward quality reproduction by loudspeaker was the invention of the moving-coil unit.

MOVING COIL

The restriction with the moving-iron type was caused by the physical stiffness of the reed. The moving-coil suspension allowed the diaphragm to move much more freely, particularly at lower frequencies, where greater movement is needed for building the bigger waves. But then designers found themselves confronting the question of how big the diaphragm ought to be.

Fig. 4-1. Before loudspeakers were invented, a way of enabling more people to hear headphones was to put them in a pudding basin.

To make a big wave, you need a big diaphragm to move all that air. But a big diaphragm gets floppy, especially when higher frequencies are used to drive it, so it breaks up at higher frequencies. This means that part of the diaphragm moves back when the part driven moves forward, and the upper frequencies get mixed up; some are reinforced, while some are almost canceled (Fig. 4-2).

Making the diaphragm smaller, enables it to handle the higher frequencies better, but then it is not big enough to push all the air to make the big waves needed for the lower frequencies. How could the whole range be handled better?

That discusses the rate of movement based on size. Another feature that limits a loudspeaker's capability and quality is the amount of movement—how far the cone has to move, particularly at the low frequencies to get the sound out. At low frequencies, enough air to make the waves can be moved either by a large diaphragm or cone moving a relatively small distance, or by a small cone moving a much greater distance.

But without the baffle, which came later, it is not a simple question of size. For example, a cone 2 feet in diameter has an area of a little over 3 square feet. A cone 6 inches in diameter has an area of a little over 3/16 of a square foot. The big cone has a diameter four times that of the small one, sixteen times its area.

Now at first sight it might appear that, by making the small one move 16 times as far as the large one, it would move the same amount of air and produce as much sound wave as the large one. But this is not true at the low frequencies. Even if it was, what would it mean? If the big one moves 1/100 of an inch, the small one would have to move 16/100, or about 1/6 of an inch to produce the same total air movement.

But air moved by the big speaker has to go a long way to get from front to back, or back to front. The path around the edge of the smaller one is much shorter. So to get the same sound wave generated, the small cone would need to move perhaps more than half an inch, which is too much for most loudspeakers.

This was what led to the next step, the invention of baffles.

BAFFLES

Quite early in the game, the baffle principle was discovered. A smaller diaphragm will not radiate the big, low-frequency waves, because air escapes round the diaphragm's edges (Fig. 4-3). It stirs the air at these frequencies instead of radiating sound waves.

The problem with this circulation of air round the edges of the cone is that you get a lot of air movement with very little sound wave going out into the air. The diaphragm or cone has to move a long way to get that air movement, but you have little sound wave as a result.

Putting the loudspeaker unit in a baffle, which was simply a large piece of plywood in which a suitable hole was cut, helped the

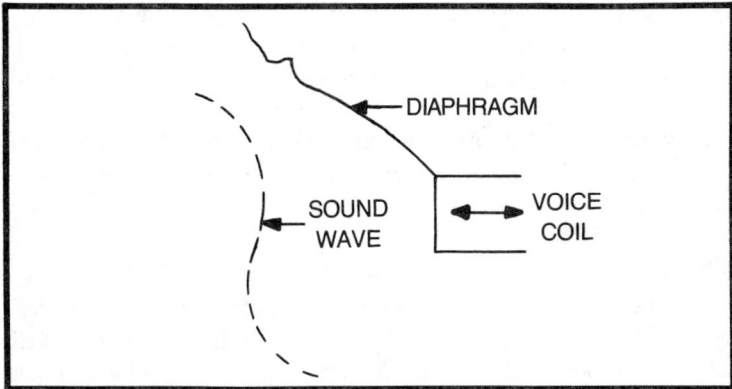

Fig. 4-2. The disadvantage of a large loudspeaker unit, is that the diaphragm almost inevitably gets floppy at the higher frequencies, producing an erratic response in this region.

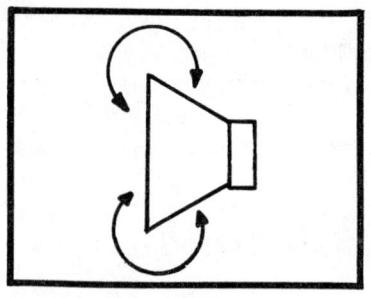

Fig. 4-3. The disadvantage of a small loudspeaker unit is that it cannot get the big wave, low frequencies out because the waves shuffle round the edges of the diaphragm.

low frequencies enormously by stopping this circulation round the edges (Fig. 4-4). The improvement that a simple baffle can give to bass has to be heard to be believed, even today.

This depends to some extent on the size of the cone still. But in a baffle, the foot-radius cone will move sixteen times as much air out as a sound wave as the cone with 3-inch radius, provided the baffle is the same size outside in each case. Or, as we put it just now, the 6-inch cone will move as much sound wave by moving sixteen times as far as the 2-foot cone.

But the limit to low frequency response, by using a baffle is now transferred from the size of cone used to become the size of baffle used. To make it easier to see what happens, imagine a really big cone, say 4 feet in diameter. Now replace that with a round baffle (although they are usually square, but we want to illustrate a point) that is 4 feet in diameter.

If you put a unit with a cone of 2 feet in diameter in this baffle, it will need to move four times as far to produce the same sound wave as the 4-foot cone with which we started. Now, if you have a cone 1 foot in diameter, it will need to move sixteen times as far, to produce the same sound wave. And, if you come down to a 6-inch cone, it will need to move sixty-four times as far, to produce the same sound wave.

But even at that, each of the smaller units will sound much better in the baffle, as far as low frequencies or bass is concerned, than out of it.

HORNS

But even before the development of baffles, musically minded designers had applied a principle that manufacturers of horns of all sizes had known for centuries. The quality of a horn's tone depends on the flare rate of the horn, and the size of the horns bell, or mouth. But musical horns are tuned to play the variety of musical notes of which each horn is capable.

Mathematicians accustomed to such calculations figured that making the horn exponential in its flare rate would make it handle all frequencies uniformly, and at the same time produce efficient transfer of energy from a relatively small moving diaphragm, first to a narrow column of air at the throat of the horn, then to a larger volume of surrounding air at the horn's mouth.

How does a horn do its job? The most primitive horns were the old megaphones that actors and coaches at one time used. Some coaches still do. They speak into the small end, and their apparently amplified voice comes out the large end. How does such a horn manage to amplify a person's voice?

Partly, it happens by directing the person's voice, thus concentrating it instead of wasting it in unwanted directions. But is that all? If a person shouts, without a megaphone, how widely does his voice spread? Not that much wider, is it? Part of the reason for the improvement is that it conserves some of the vocal energy that comes out of his mouth.

A similar thing happens, perhaps even more effectively, when a horn is used on a loudspeaker unit. Even with a baffle, a loudspeaker diaphragm, or cone, has to move air. Because the paper, or whatever the cone is made of, is so much heavier or denser than the air it has to drive to make a sound wave, it spends most of the energy fed to it in driving the cone, and only a small part in producing the sound wave in the air.

If we can find a way of making the air seem heavier to the cone, we can use more of the energy in moving that air. That is what a horn does. But it must also deliver the sound wave, eventually, into the air, which is no heavier in fact, than normal. That is what the horn also does.

At the throat, where the column of air is narrow, the cone or other shaped diaphragm driven by the voice coil (often it is domed, rather than conical) is invariably larger in diameter than the column of air that it drives. This has the effect of making a small movement of the diaphragm produce a larger movement of air in the throat.

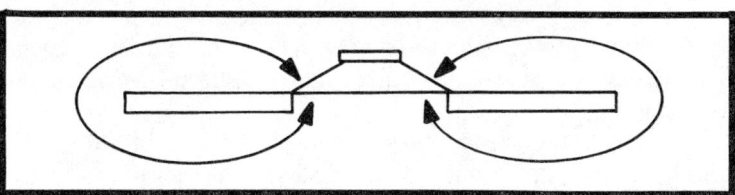

Fig. 4-4. Putting a loudspeaker into a baffle does wonders to help the low frequencies by stopping the shuffling round the edges, or at least making it go further.

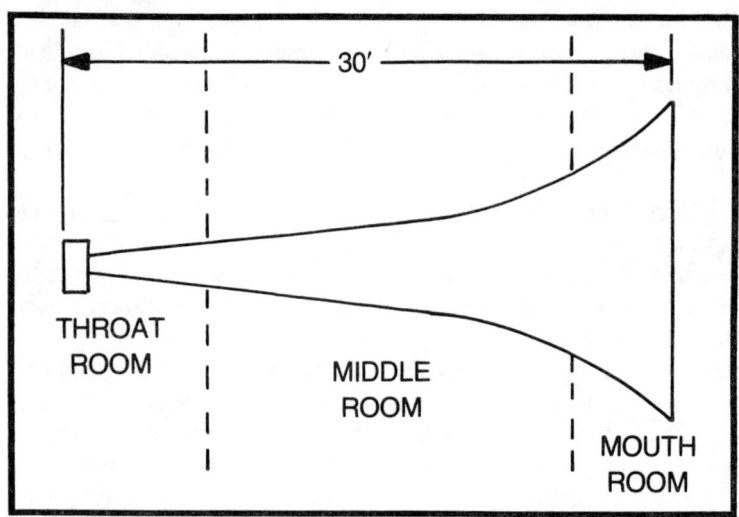

Fig. 4-5. How the big horn at the British Science Museum is housed going through three rooms. How big of a house would you need for quadraphonic, at this rate?

That larger movement is accompanied by larger pressure fluctuations. Then, as the wave travels down the horn, expanding, both movement and pressure reduce, until the wave is ready to move out into the free, open air. The overall effect of a properly designed horn, is to efficiently couple the diaphragm so it puts much more of its total energy into the air at the throat of the horn, and then expands it without further loss, until a fully developed sound wave emerges at the mouth end.

The exponential horn at the British Science Museum in London, is more than 30 feet long, and its mouth is several feet across. It was installed there before many of the modern types we shall describe in this book were invented. But it is a very big loudspeaker, whose pressure drive unit is in one room, the body of the horn goes right through another room, and the mouth where the sound comes out is in yet a third room (Fig. 4-5).

That is obviously not a practical arrangement for most of us, especially if we want stereo, or even quadraphonic reproduction. That is the main reason for other kinds of loudspeaker being invented.

However, the science museum horn can be scaled down in size, which limits its performance, as we shall see. A popular size uses a mouth not much more than a foot across, with a length of 3 or 4 feet. Many such horns were used, in the earlier days of sound reproduction, and they still have their uses.

Such a horn will handle frequencies above a cut off frequency that is determined by two things: the size of its mouth, and its flare rate. For voice, such a horn is good, but for music, because it handles nothing in the middle register, only the upper frequencies, it sounds thin and tinny.

A horn is, without doubt, the most efficient kind of loudspeaker. That 50% efficiency, achieved by very special design, used a horn. So, in the days when acquiring electrical watts output was quite expensive and involved large and costly vacuum tubes, horns did provide the best solution. In those days, to overcome the size objection, designers folded the horns into a cabinet size that was not unreasonable for a fairly large living room (Fig. 4-6).

By placing such a horn in one corner of a room, where two walls and either the floor or the ceiling meet, usually on the floor, the actual corner of the room, that consists of three flat surfaces fanning out at 90-degree angles, forms the mouth of the horn. Thus, the effective size of the loudspeaker is quite a bit bigger than the actual box that houses it.

Thus the loudspeaker must be regarded as part of the room, or the room as part of the loudspeaker, whichever way you prefer,

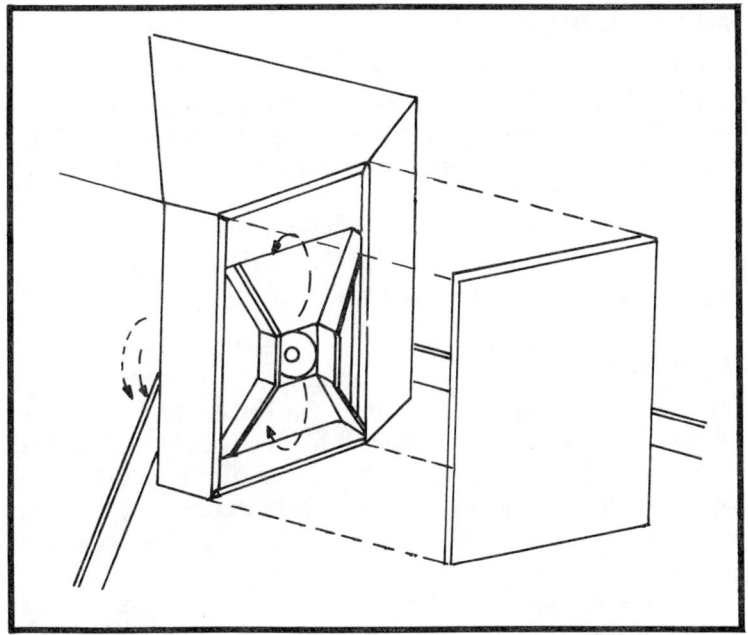

Fig. 4-6. One arrangement for a folded horn with the front panel removed, to show some of the inside development. Dashed lines show the path taken by the sound, as the wave expands.

when you use a folded horn. Corner horns enjoyed quite a long period of popularity among those who could afford them as *the best*, which they were, in the days of monophonic high fidelity.

For a single loudspeaker, you needed only one such corner in a room, and the sound it delivered was efficiently converted from the output of the much smaller amplifiers that were available in those days (smaller in power output, but much bigger physically). When stereo came in, to use horn loudspeakers you needed two such corners, at opposite ends of one wall of the room. That is, they had to be in adjacent corners, not opposite corners of a room.

Not all rooms were that convenient, even where they were big enough to accommodate two such large pieces of furniture. So stereo really reduced the popularity of horn loudspeakers. However, they had started to decline before the advent of stereo, even when they were the best for rooms and people that could afford them.

But smaller houses and apartments were becoming fashionable for homes, and the large horn cabinet is totally impractical for such locations. So designers sought and invented other ways to get results.

PRACTICAL BAFFLES

Although the baffle was less efficient than the horn, it was also much less bulky. It did involve a large, flat board, and the bigger the better. Those who had the ambition to use a built-in loudspeaker, would cut a hole in the wall of their living room, and mount the loudspeaker unit in the hole, which was very effective (Fig. 4-7).

But what do you do when you move out of a house where you have done that? If you take your loudspeaker out, maybe the new people will not want a hole in the wall!

THE INFINITE BAFFLE

These problems with the simple flat baffle board led to the design of what was called the infinite baffle. They had found out that low-frequency response was limited by the size (or smallness) of the baffle. And the reason for this was the fact that increasing path length from back to front stopped the air-shuffling effect round the edges down to a lower frequency.

So the idea that led to the infinite baffle, and to giving it that name, was that if you box in the back of the loudspeaker, air can never get from back to front, so the size of the baffle is virtually infinite.

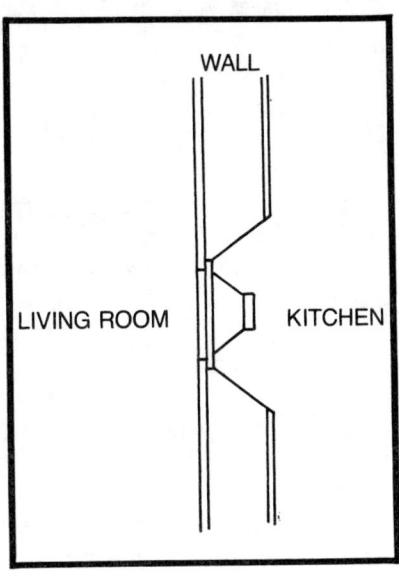

Fig. 4-7. Using a wall as a baffle for a built-in loudspeaker. A wonderful idea, but do you sell the loudspeaker as part of the house?

That theory was very nice but, as we see later, this too had its limitations. As with so many other things, you find you can never get something for nothing.

EFFECT OF RESONANCE

Boxing in the back led to discovery of another limitation. This concerns the resonance of a loudspeaker unit. However large a baffle is used, the loudspeaker's low-frequency response is also limited by the diaphragm's mechanical drive system. This centers around a frequency known as resonance, at which, under the proper conditions, the compliance of the diaphragm suspension, working with the weight of the diaphragm and the air in contact with it, twangs like a spring with a weight.

To understand this, at frequencies above resonance the diaphragm must move its own mass, or weight, along with the weight of air against it; the stiffness of the suspension has very little effect. At frequencies below resonance, the diaphragm uses most of its effort pushing against the stiffness of the suspension, and very little is left for pushing the wieght of the diaphragm and the air against it.

As moving the air against the diaphragm is what makes a sound wave, any loudspeaker unit radiates very little below its resonant frequency, being retarded, more and more, by the control exerted by the stiffness holding the diaphragm.

Let us just review that. For the moment we are not talking about what the box does, although that does affect it as we shall see in a moment. We are talking about where the energy, supplied by the amplifier and fed to the voice coil, goes at various frequencies. At resonance, which is where the mass of the diaphragm resonates with the compliance of its suspension, most of the energy goes into providing this big movement.

But on either side of this frequency, above and below it, movement reduces for the same driving force or electrical output from the amplifier. Why? As we said, at frequencies below resonance, the major opposition to diaphragm movement is the stiffness of the suspension, however that is made up. So most of the energy goes into flexing that suspension, little into pushing the actual diaphragm or the air in contact with it.

At frequencies above resonance, the major opposition to movement is the mass of the diaphragm and the air it has to push with it. As pushing the air is what makes the sound wave, this is where the loudspeaker realizes whatever its maximum efficiency is.

That is generally true about resonance and frequencies above and below resonance, whatever it is within the unit itself that determines at what frequency resonance occurs. This is where the kind of box or enclosure comes into the picture, because it always affects the resonance frequency among other things.

BOX SIZE

So how does boxing in the back affect this? Because, when the back of the diaphragm is enclosed, its movement alternately compresses and expands the air inside the box, which cannot get out. This is exactly like adding more stiffness to the suspension. It has the effect of pushing the loudspeaker's resonance to a higher frequency. Let us step through what baffling does, one step at a time.

If a 10-inch or 12-inch unit has an unmounted resonance, which means just the unit held in space with nothing round it, at about 90 hertz, which would be typical, then putting it on a large open baffle will lower that resonant frequency by making it move more air, which means the weight the diaphragm has to move is greater. Perhaps the resonance will move down to about 60 hertz (Fig. 4-8).

Notice that two things happen here. The resonance frequency is moved down from 90 hertz to 60 hertz, but also, because more air is moved, the sound output is increased much more than just changing the resonance would indicate.

Now, if the same unit is put in a box, first with the back open, then with it closed, see what happens (Fig. 4-9). Putting it in the open-backed box has much the same effect as putting it in a baffle. But now, when you put the back in the box, the resonance goes even

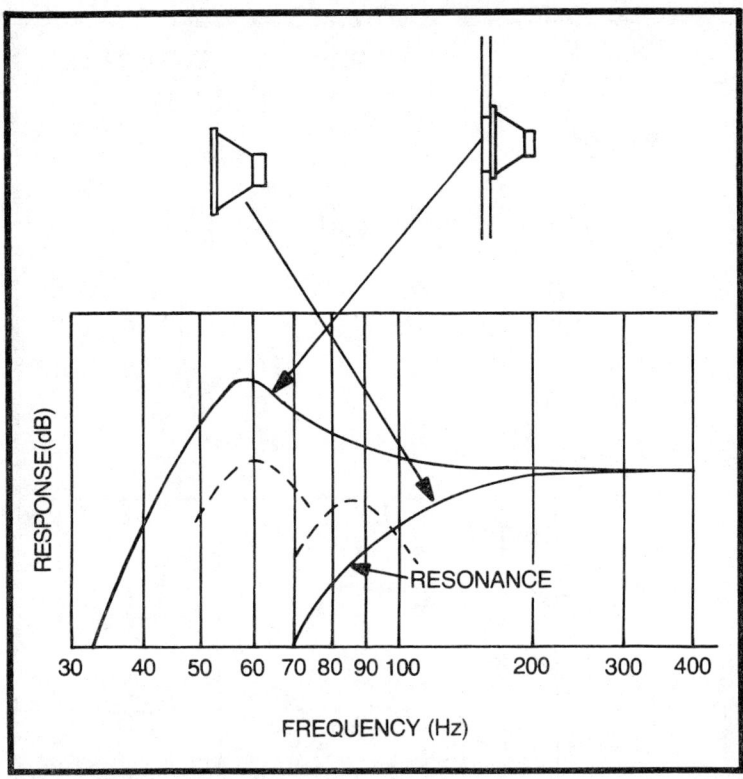

Fig. 4-8. What putting an ordinary loudspeaker unit in a baffle does to its resonance and the response.

higher than it was at first, up to say 150 hertz, depending on the size of the box.

Now, you are making the unit do more work, compressing and expanding the air inside the box, than it does radiating the sound wave out front, so it loses efficiency, represented by the drop in level when the back is put on. And the smaller the box, the higher the frequency that the resonance moves to, and the more efficiency is lost.

The original stiffness, due to the physical suspension of the loudspeaker, does not change. Let out of its box, the unit would perform the same. But the effort of compressing and expanding the air inside the box adds more control to the diaphragm movement, making its action stiffer.

As well as shifting the resonance frequency upwards, this causes more energy to be used in compressing and expanding the air in the box, below that higher resonance frequency.

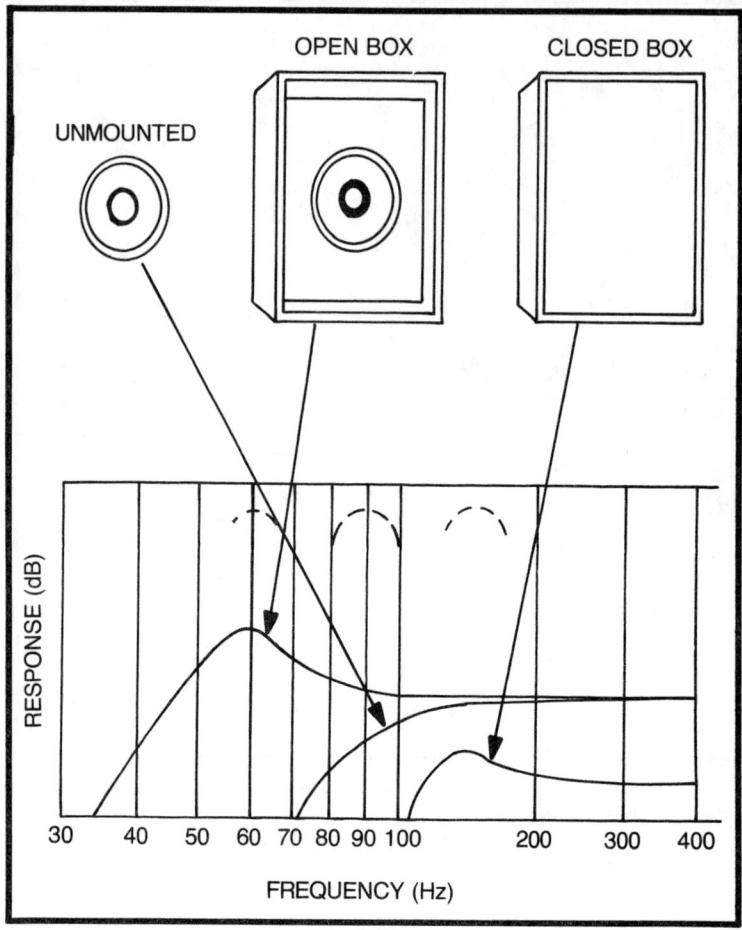

Fig. 4-9. The effect of going from an open-back box, which is much like a baffle, to a closed-back box, step by step.

So using an infinite baffle does not gain the advantages promised, by keeping in the sound from the back. One more thing: an infinite baffle, as well as later types that modify the same principle, rely on having the back sealed. A break in that seal will change the performance.

BASS REFLEX

But what if we deliberately let the air escape from behind, through an opening specially designed for the purpose? This idea was the basis for the bass reflex. In the classic bass reflex, the hole adjoins the opening for the front of the diaphragm (Fig. 4-10).

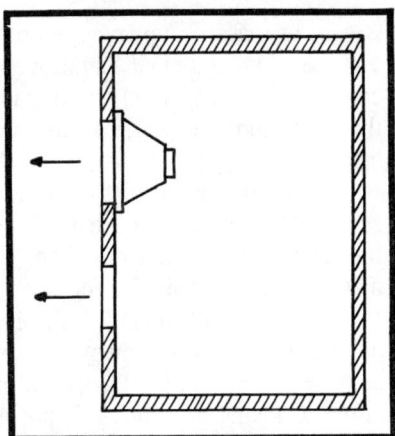

Fig. 4-10. The idea that became the classic bass reflex loudspeaker.

To help understand how the bass reflex works, let us take a look at a form of loaded reflex first (Fig. 4-11). This has a dummy diaphragm mounted in a second hole, exactly like the one the main loudspeaker has, except that the dummy has no electrical drive. Because both diaphragms are the same size and weight, resonance occurs when they both move together, and the air inside is compressed and expanded by both of them.

At resonance frequency, quite a small amount of energy fed to the driven diaphragm will maintain both of them moving in this fashion. The simple bass reflex substitutes a hole, where the mass of air in and around the hole achieves the same effect without having to move an actual diaphragm in the hole.

It will help at this point to think this through a little further. When the second hole in the box is occupied by an undriven cone, the

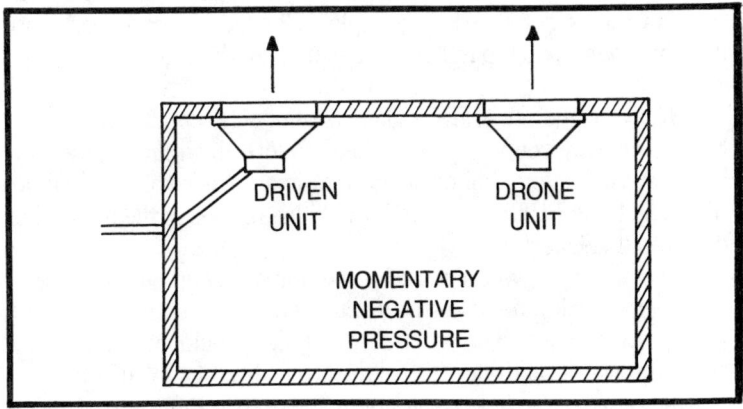

Fig. 4-11. Understanding how a bass reflex turns the back wave round.

weight or mass occupying the hole consists of the cone, plus the equivalent amount of air that moves with it. Now, if you remove the cone, leaving an open hole, the mass will be reduced by the mass of the cone, and thus will not tune at the right point, but at a higher frequency.

How do you bring this back so the air hole reinforces the reproduction at the lower frequency as it did when the cone was in there? To answer that question, we must ask another. What will make the mass of air moving in the hole seem bigger?

If you think making the hole bigger will do that, you would finish up taking the whole box away before you got there. As we shall show in greater detail later in this book, mass, when we are talking of the effect of moving air, is not a simple quantity. It must be related to the space in which it moves.

What we are thinking about resembles a Helmholtz resonator, of which a simple example is an ordinary bottle. Have you ever tried to see how the size of the bottle, and the size of the neck, affects the frequency at which the bottle resonates? You can make such a bottle produce its resonant frequency, by blowing across its mouth. You have to get your angle of blow just right, and blow fairly gently—not a powerful blast—and the bottle will build up a note that is its resonant frequency.

First, you will find that the larger the bottle, the lower the note, if its neck is the same size. In this connection, you will find that bottles with smaller necks, relative to their volume, are easier to make resonate. This is important to what we are talking about, too.

Next, if you get, say two 1-gallon bottles, with necks of different size, which gives the lower note? You will find it is the bottle with the smaller neck. The length of neck makes some difference too, but this is not so easy to explore, because longer necks are usually tapered, not parallel, which means their diameter is not constant.

If you happen to know a glass blower who could turn you out bottles any way you want, you could explore this much more fully, but as most readers will not have this luxury available we will not pursue it. So we will get back to how all this relates to the bass reflex design of loudspeaker.

It means that, when you take the dummy cone out of the hole, you need to make the hole smaller, to bring the resonant frequency to the same point. The area of the hole should be somewhere between one-half and one-third of the area when occupied by a cone.

If you make the hole smaller yet, the resonant frequency will go even lower. But because the hole is smaller, it cannot radiate so much sound wave. And it will not match the equivalent mass of the speaker cone, which occupies the other hole. So you will find it will not work very well. But by proper tuning, making the hole the right size, a bass reflex box lowers the resonance and reinforces its sound output at that frequency.

A bass reflex loudspeaker extends the response to a lower frequency than the same sized unit mounted in an infinite baffle of the same size because it can move more air that goes into the outside world.

LOW-FREQUENCY DIFFERENCES

So we have three basic types of loudspeaker, omitting the simple baffle for the moment, which we will back to later. These are the horn, the infinite baffle, and the bass reflex. The main difference in how they work is at the low-frequency end. The horn is the biggest, for a given low-frequency response, the infinite baffle comes next, and the bass reflex is the smallest.

HIGH-FREQUENCY DIFFERENCES

For high-frequency response, all three types tend to get directional. They squirt the higher frequencies along the axis of the unit, unless something is done to change it. Horn units can be made multicellular, which splits the high-frequency tone pushed along the horn into several paths that are centered in different directions (Fig. 4-12), enabling directionality to be tailored to a system's needs rather precisely.

But multicellular horns can be used only for higher frequencies, above their cutoff, which is usually around 1000 hertz. To diffuse higher frequency sound from the diaphragms of units in the other type, called direct radiators, a variety of methods have been devised, none of which is perfect, but all of which are better than nothing, where diffusion is needed for effective reproduction.

MULTIWAY SYSTEMS

Because handling the different audible frequencies calls for various design features, none of which apply to the whole frequency range, many modern loudspeakers are built as two-way or even three-way systems. One unit handles the lowest, bass frequencies. Another unit handles the highest, treble frequencies. And perhaps a third unit, called the midrange, handles a range sandwiched between the two extremes.

Fairly obviously, multiway units cost more than simple units. And when they are used, proper combination is essential, something we cover in Chapters 10 and 11. Here we need to answer some questions about whether multiway or single unit systems are best for a given application.

From the viewpoint we are discussing now, the question is which will do the job best, as far as converting the program to be heard from its electrical to its acoustical form. As the signal comes through the amplifiers, in one, two, or four channels, each is a composite of audio frequencies, varying in time—a complicated waveform of voltage and current.

Reverting to the ideal that was envisaged before various complicating complexities were encountered, a good loudspeaker should produce an acoustic wave that accurately translates the electrical input wave. As we have already seen, the real need is much more complicated than that.

As well as having component frequencies faithfully reproduced with acceptably low distortion, the loudspeaker must distribute the acoustic waves in a proper manner appropriate to the listening environment. This is the problem that loudspeaker designers have been addressing continuously for the past 50 years. Like many other branches of technology, significant advances have been made toward solutions to the problem, an understanding of which cannot be learned in a few words.

Most often, the main question reduces to the problem of getting better bass from smaller boxes. Getting the response at higher, treble frequencies, is not usually so much of a problem. What problem there once was, mainly required greater precision in production techniques, and technology has found those remarkably well.

High-frequency deficiencies in response were largely due to quite small variations in shape or to tiny cavities, where something like a high-frequency Helmholtz resonator might be simulated. For example, in a high-frequency horn compression driver, earlier designs might have a cavity under the dome (Fig. 4-13).

With the diaphragm, this makes a resonator that produces a kinky response well within the audible range. The better way, also shown at Fig. 4-13, is to have diaphragm almost touching all over on one side with a number of communicating ducts to allow air movement to get out. Technology for making such designs with high precision has improved performance of such loudspeaker units.

Making the voice coil and diaphragm to very close tolerances in shape, and reducing the clearance between the coil and the magnetic pole pieces, has all helped to produce better and more consistent units for high-frequency performance.

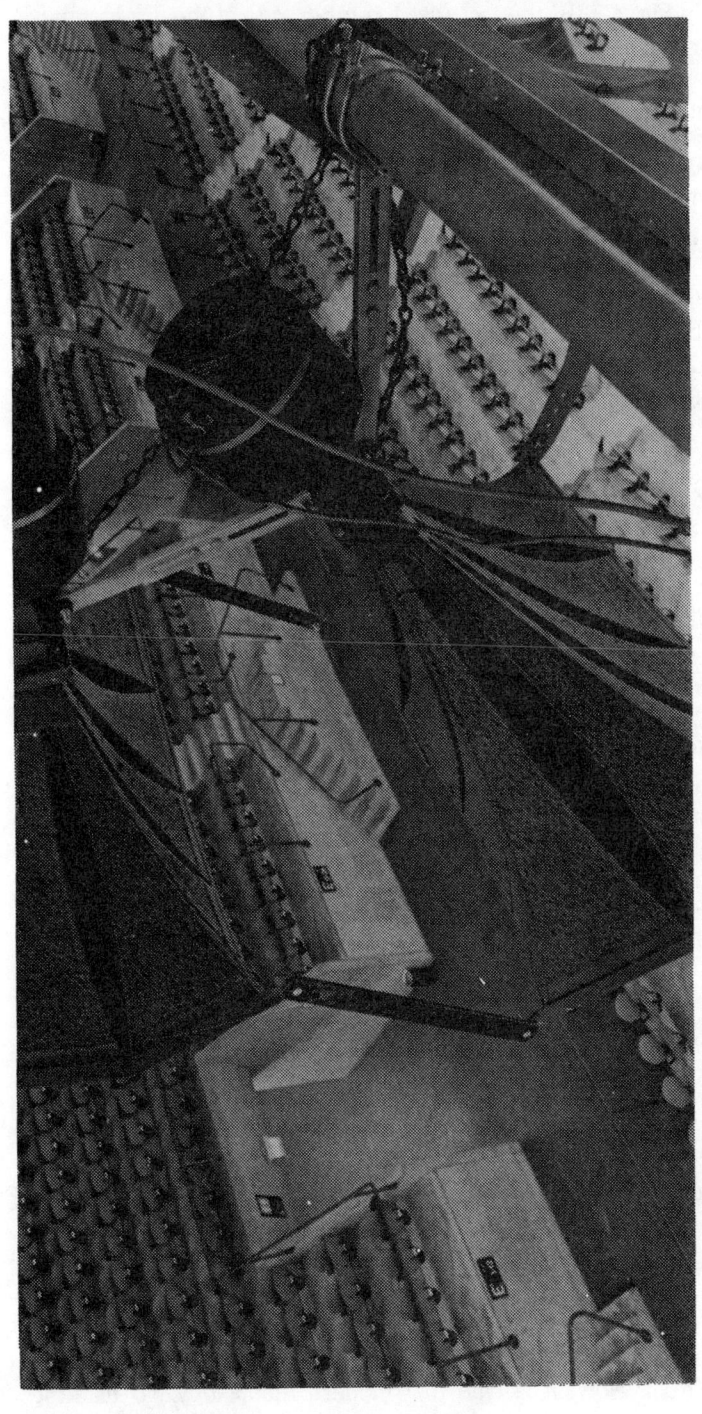

Fig. 4-12. An example of multicellular horn use, to direct high-frequency sound at a controlled area of seating. This picture was taken at the installation used at the Montreal Olympic Games. (Courtesy Altec Sound Systems.)

In Fig. 4-13B, the fact that the other side of the diaphragm is either completely open, or covered with something acoustically transparent so that no effective cavity is produced may puzzle you: doesn't that let sound come from the wrong side, as well as the right side? The amount of sound from the free side is very small because the movement is restricted by the coupling to the throat of the horn.

The efficiency with which sound is coupled out through the horn may be, say 20%. In really good horns it can be higher. The efficiency of acoustic coupling to the free side is probably far below 1%, so its effect is quite unimportant.

But for better response at lower frequencies, we have already introduced the main problem: that of moving sufficient air to make the big waves needed, and of doing it efficiently without an inconveniently big box.

EFFICIENCY

It is important to recognize that all these successes do not include any way in which more bass can be obtained from a smaller box with higher efficiency. As what we need is more bass from smaller boxes, and as modern technolgoy has made electrical power at audio frequencies easier to get, what we need has had to be obtained by sacrificing efficiency.

This is done in one of two ways: one is an adaptation of the inifinite baffle, the other of the bass reflex.

We will explain what happens to efficiency more fully in Chapter 8. For now we need to understand what the objective in all these new designs was and is. The infinite baffle curtails the bass response drastically. Those who have grown up with horn loudspeakers sneer at infinite baffles and later types as terribly inefficient devices.

If all they intended to convey was a love of efficiency for its own sake, we could forget that attitude, and dismiss it as an irrelevance from the past. But one of the advantages of a well-designed horn loudspeaker, because of, or at least partly because of its efficiency, is its smooth response and clean integration of the program so everything sounds as if it belongs together. Perhaps it would be better to describe the effect as "getting the loudspeaker out of the way."

What this means, is that the sound coming out does not sound as if it is being reproduced. It sounds like the original program, as if there is no loudspeaker. Can you say that equally for any of the newer types? They may produce very pleasant sound, clean and all that. But doesn't it still sound as if it is coming from a loudspeaker?

This is something the horn-type loudspeaker managed to get away from. Now, it is true that modern technology has made it

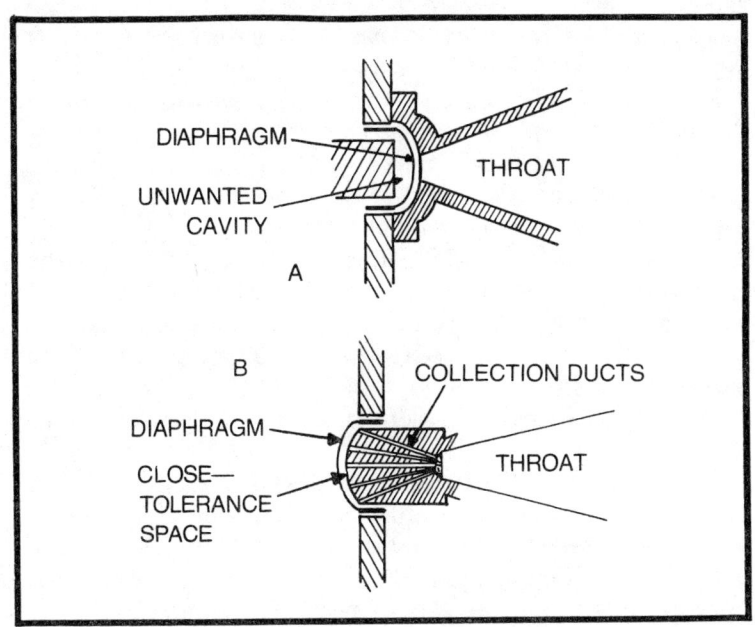

Fig. 4-13. An example of how precision technology has improved the performance of high-frequency horns. (A) An early design with a cavity that produces poor high-frequency response. (B) An improved design that eliminates the defect in example A.

possible to get 100 watts or more as easily as we used to get 10 watts or less, so that the inefficiency of the newer types can easily be made up for in the amplifiers. And very pleasing sound can result, coming from boxes that fit in with the decor, as the old horns could not in many instances.

But the horn loudspeaker has still not been superseded in this one respect. The newer types are carefully designed for as flat as possible by getting rid of various unwanted resonances, matching in the various units that combine to cover the whole frequency response and so forth. And for what they do, they do a terrific job. So now, let us move on to see how they do it.

ACOUSTIC SUSPENSION

The first designer to adapt the basic infinite baffle to give better bass from smaller boxes called his design the acoustic suspension unit. What is the real obstacle to getting big bass from a small box? Mainly that of getting the resonance of the diaphragm at a low enough frequency. The resonance frequency can be lowered in two ways: by making the diaphragm heavier, which means it will take more watts to move it; or by making its suspension softer, or floppier, so it exerts less restraint on its movement.

Using both these means at once, acoustic suspension units use thicker, heavier diaphragms and softer, more floppy suspensions. Having a heavier diaphragm makes them take more watts to move the diaphragm, leaving less watts to move the air they push. So they are inherently less efficient than other types.

Now, instead of making the suspension built into the loudspeaker unit itself the main restoring force, they rely on the air inside the box to do it. Let us look back at what an ordinary unit will do. It might have an unmounted resonance of 90 hertz. Putting it in a smaller box might shift that down to 75 hertz (see Fig. 4-9), and putting the back on such a smaller box might push the resonance up to 180 hertz.

Now, to pursue the acoustic suspension design, unmounted, the resonance might be down around 2 hertz. Putting it in the box, with the back open, would lower that, only very slightly, but it would get more output from it. Finally, putting the back on, sealing it, will bring the resonance up to 30 or 35 hertz (Fig. 4-14).

This point is that, without the air cushion effect produced by the sealed box in the back, the unit will flop back and forth at about 2 hertz. Closing the back makes the movement much more controlled. The name acoustic suspension says that what controls the diaphragm movement is the cushion of air behind it, rather than the compliance of the surround.

LOADED REFLEXES

The other main adaptation to get a much smaller bass unit that goes to a lower frequency uses a similar method, but based on an adaptation of the bass reflex design. However, because the driven diaphragm is made heavier, the port or opening must also be modified, which can be done in a variety of ways.

It may be achieved by using a tube rather than just an open hole, or it may use another heavy diaphragm in the opening. We will have more to say about this in later chapters. There are many variations.

THE FLAT RADIATOR

We should mention one more variation used in modern installations. This reverts to the original flat baffle idea but with some basic differences in application. In its modern application, it generally uses a slim-line speaker unit, so the overall structure is quite flat and not very large.

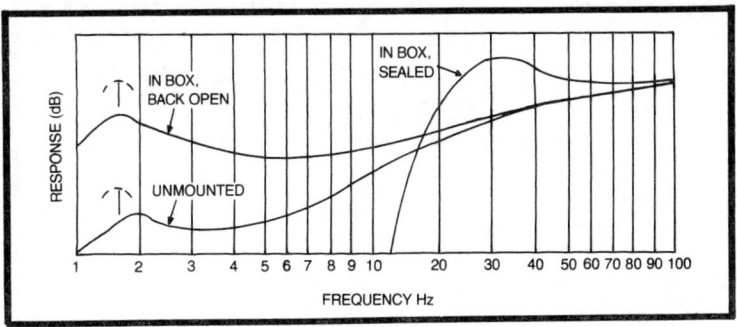

Fig. 4-14. How the response of the acoustic suspension type is achieved.

It achieves its effect by utilizing radiation from both front and back, a little differently from the earlier type baffles, at which time not enough was known to make use of this particular effect. In the live type room, another way to avoid the effect of the many spurious reflections is to give the listener's hearing something better to work with, as a replacement for the direct wave to which he normally listens.

The advantage of an open-back loudspeaker is that, within fairly short distances (characteristic of this kind of room) from the unit, it creates a complex sound field, not just straight radiation like the closed-back one does. This complex field can then create a field round the individual listener's head that gives him an illusion you might not expect.

The principle can be shown to work mathematically, but hearing it work for yourself under the circumstances it suits is far more convincing than a set of mathematical formulas.

Figure 4-14 shows a typical arrangement that works well with stereo in that type of room. The lower frequencies that the relatively small baffle type units do not radiate well and that have big waves are channeled together (from both left and right stereo channels) into a single bass unit.

Then the baffle units carry everything above these low frequencies. The changeover may be anywhere between about 100 hertz and 300 hertz.

Without belaboring the explanation, the extraordinary thing is that, particularly in that kind of room, all of the sounds seem to come from a stereo left and right (or some intermediate, where that is appropriate) and the apparent sources are not even located where the flat radiators are actually placed

Wherever the listener is located in a room served by a system like this, the sound field round his head is somewhat like the one at

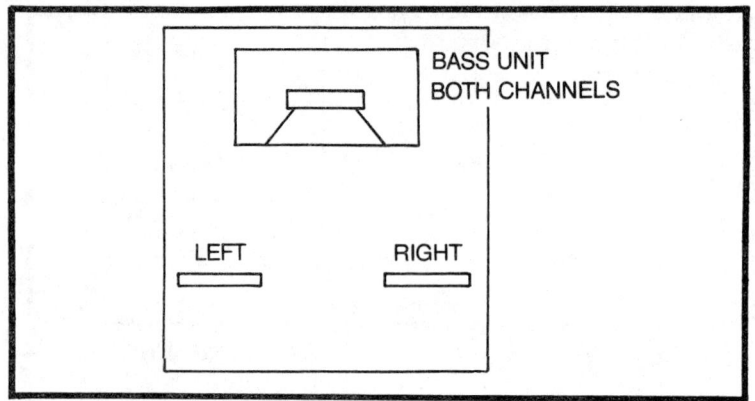

Fig. 4-15. A typical stereo installation for a live room that uses flat open-back units for left and right channels. This uses an advanced technology with some unexpectedly good results.

the edge of a baffle board, or as we shall describe later (Chapter 10) as a result of two loudspeakers being connected out of phase. But doing it in this way, the sound field is controlled, so it gives a desired impression to the listener, wherever he is, produced by the combined effect of both these open-backed units.

This system is something that has to be heard to be believed. You may thing there is something synthetic about it, but really no more so than any of the newer types of loudspeaker produce. And the way the sound melds, to produce an illusion is unbelievable. For example, suppose part of the program consists of an old-fashioned string bass played somewhere off-center (it does not matter where, exactly).

The body sound of that string bass must be coming from the single low-frequency unit, wherever that is located. The other units just do not handle it—it is not fed to them. But such things as the pluck sound and the overtones come from the other units. And the effect you get when listening is a definite location of that instrument, wherever the stereo recording or programming meant it to be quite precisely.

Although you know, if you understand how the system works, that the main low-frequency sounds that come through with such lovely body must be coming from somewhere else, listen as you will, you cannot detect the fact audibly.

What happens is a sort of acoustic synthesis that works very well, allowing the listener more easily to ignore the room-generated reverberation that so easily swamps out the sound from the more conventional closed-back loudspeaker in this environment.

5

Picking The Right Kind For The Job

Now that we have covered the factors about the choice, and the kinds we have to pick from, we can proceed to match up the two in order to pick the right kind for any particular job. What are the factors? Room size, quality, kind of program choice, and the way the program is to be delivered, such as mono, stereo, or quadraphonic.

HANDLING THE FREQUENCIES

And what are the choices? For bass, or low-frequency reproduction, you will not often want a horn, although there are still uses for them, which we should cover. For most situations, your low-frequency response is cared for with a choice between infinite baffle and bass reflex, or between modifications of them, such as acoustic suspension and loaded reflex (Fig. 5-1).

You want to pick the right kind of unit for reproducing this end of the frequency spectrum based on the factors we have already discussed. The only type of loudspeaker that truly integrates low-frequency sounds with high-frequency components of the same sound so well that it does not sound as if you are listening to a loudspeaker is the horn. So if space and the room layout permits that is your best choice, cost also being a factor possibly.

But assuming that is out of the question, you have the choice of various other methods of getting bass output. The thing to realize is that you will not locate sound source by the unit that produces the lowest frequencies, so you need to pick a unit that gives you the best size compromise for the room you have as well as "getting down there" as well as possible.

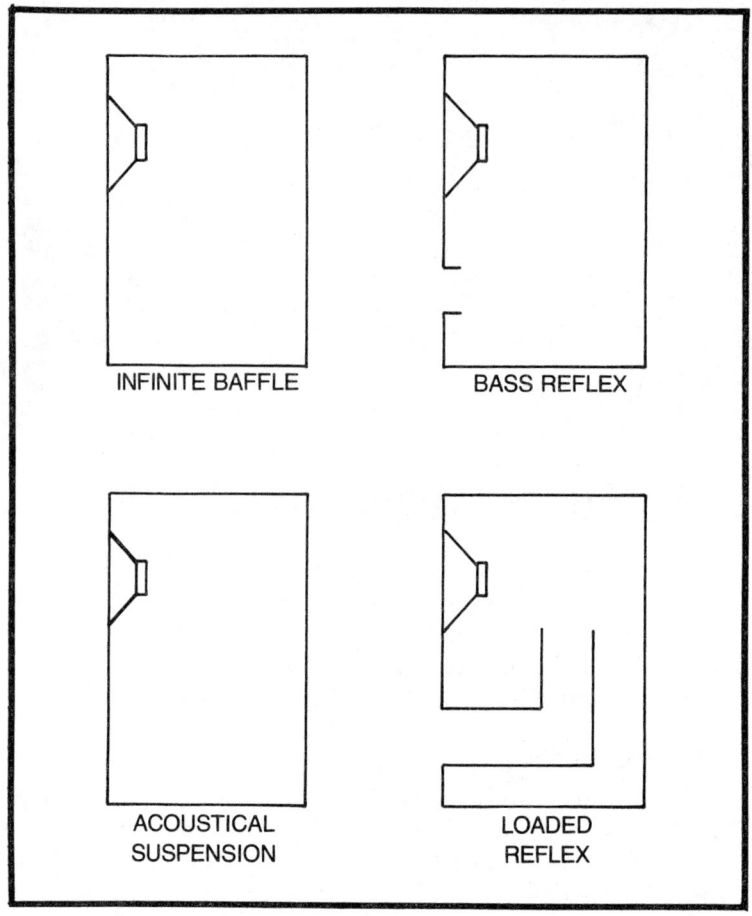

Fig. 5-1. The most common choices of loudspeaker type, for handling the lowest frequencies.

One thing you need to think of, is the ability of the unit to fill the room adequately with sound of these low frequencies, which will vary with type. As we have said, in general, the power output capability at the low-frequency end is roughly determined by the size box it comes in, although there are the differences in capability for size that we have already discussed.

For high-frequency performance, there are individual diaphragm units, called "direct radiators" (Fig. 5-2). All of them depend to some extent on the material of which the diaphragm is made, which can deteriorate with age. Finding materials that do not deteriorate has been a loudspeaker designer's nightmare, but nowadays there are some good ones.

This end is where the kind of differences we introduced in Chapter 1 can have the most effect. For this reason, when auditioning loudspeakers at an "audio salon," you should carefully observe the qualities of the room in which you listen to them. It should be as closely similar to the room in which you want to install them as possible, otherwise your listening is extremely apt is mislead you.

Note particularly the size of the room. It should not be too much larger or smaller than the one you plan to install the loudspeaker in. As we have pointed out, a loudspeaker that sounds well in a smaller room, can sound quite inadequate in a larger one. Or one that sounds well in a larger room is incapable of giving the more intimate performance a smaller room needs.

And note also the quality of the room. How absorbent (dead) or reflective (live) is it? We have already discussed the difference that this makes at some length. The point is to be aware of it, and to listen with that in mind. If you cannot get a room that is close to being identical with the one in which you will install the loudspeakers you eventually purchase, by being aware of the differences you may be able to audition intelligently to allow for the differences.

DIRECTIVITY QUESTIONS

Where you need a directional type, look for that feature in spec sheets and listen for it when you audition the unit. Where you need a type that diffuses the sound well, it must have been designed to do that. A horn can use special shaping with surfaces that bounce the sound around before it leaves the unit, or it can use an acoustic lens built onto the unit. These are two ways of achieving this. (Fig. 5-3).

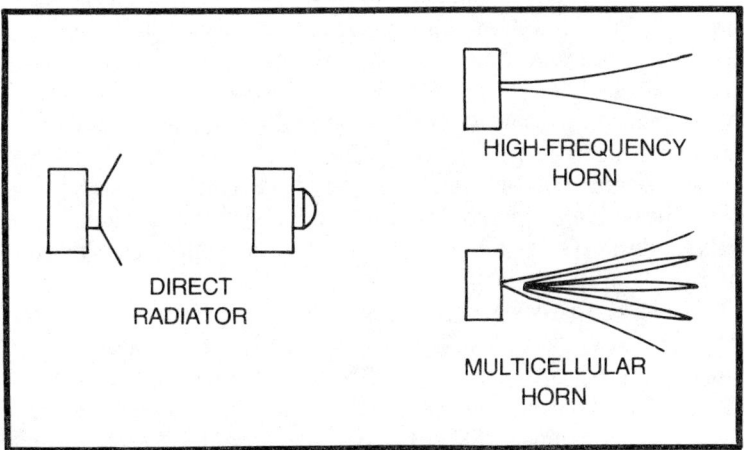

Fig. 5-2. Some of the choices for high frequency radiation.

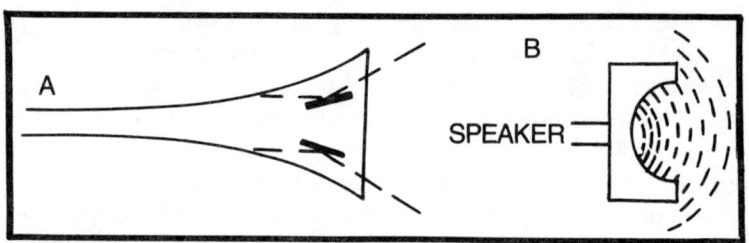

Fig. 5-3. Ways of diffusing the high frequency response.

Which one is best in a given instance depends on the needs of that case. The acoustic lens gives the most uniform dispersion of the high-frequency as well as low-frequency sounds. The baffle plates bounce the main beam into roughly defined segments so that uniformity of distribution at the high frequencies is not so good. However, for some purposes that may be adequate, or even desirable.

The linkage between loudspeaker size, and the size of room in which you put it, affects mainly the bass, or low-frequency response, as we have said. But here you must regard the room as part of the acoustic system into which the loudspeaker diaphragm radiates. This recognition is more important with some types than with others.

The folded corner horn, that was more popular when larger rooms were popular and high fidelity meant mono, must be put in the corner of a room. This is because the two walls and the floor, or if you put it at the top of the room, the two walls and the ceiling, form the final flare (Fig. 5-4). Any other type of loudspeaker gives you a little more option in where you can put it.

But you will always find that where you put the loudspeaker will make a difference to how the bass or low-frequency part sounds. The worst place to put a loudspeaker, for reproducing bass, is on a small table, elevated above the floor, in the middle of the room (Fig. 5-5A). It is better on a large table, on the floor against the wall, or best yet in a corner where two or three surfaces meet.

These remarks apply not to the very lowest audio frequencies, which have wavelengths so long that where you put the loudspeaker makes little difference, but to the lower register in music, depending on the size of the room.

As we said earlier, where physical size takes effect is subject to the dimensions of various objects, particularly the room in which sound is reproduced and the sized boxes from which it is reproduced. This range usually corresponds with more or less of the lower register in music. Thus, taking a living room 8 feet high with dimensions of 16 feet by 20 feet, an average dimension is about 13½

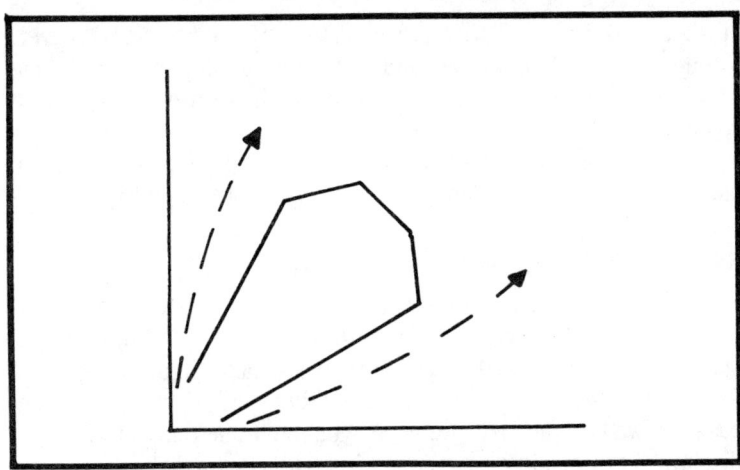

Fig. 5-4. Why the corner position is vital to good reproduction from a folded corner horn.

feet, which is a wavelength at 80 hertz or a half-wavelength at about 40 hertz.

For notes below that frequency in that sized room, if the doors and windows are closed the room acts as a pressure box. If the room is smaller, the frequency below which this happens is higher than 40 hertz. For notes above this frequency, and up to a frequency fixed by the size of the loudspeaker box, we have to think about how the loudspeaker gets its sound waves out into the room.

If the box has its major dimensions about a foot each way, this is a half-wavelength at about 500 hertz. It is one sixth of a wavelength

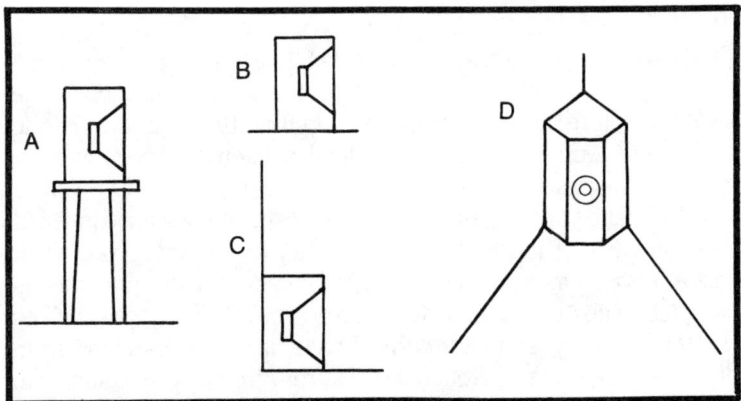

Fig. 5-5. How placement of most ordinary loudspeakers affects bass response (see text): (a) on a small table; (b) on the floor, or a large table; (c) on the floor against a wall; (d) in a corner.

at about 170 hertz. The range between 40 hertz, or whatever frequency is set by room size, and 250 hertz, or whatever frequency is set by loudspeaker box size, corresponds with what is usually regarded as the lower register in music.

Being aware of this relationship will help you to picture what is happening, or what you can expect to happen, through this range.

SINGLE UNIT OR MULTIWAY?

Probably the place to start in choosing a complete loudspeaker is to decide how you will go about getting your bass, or low frequencies. If you can use a broad range loudspeaker that radiates all frequencies, then you do not need a multiway unit. But if, to get the bass you want in the size you want, you must use one of the floppy diaphragm types, either acoustic suspension or loaded reflex, then the bass unit will not be able to produce your high frequencies as well, and you must have a multiway system.

If you are buying a ready-made loudspeaker or matched sets of them, they come with tweeters, or midrange and tweeters, whenever the bass unit uses either the acoustic suspension or the loaded reflex principle. But if you plan on building your own, you need to either buy a kit, consisting of the units you need, or to pick them out for yourself.

Either way, you need to be sure that the combination suits the room in which you want to use them. But the first decision to make is usually whether your situation requires single units or multiway types.

INTEGRAL SOUND

Having made that decision, the next thing is to get a system that is well integrated so the sounds that come out of it all belong to one another. Before the days of stereo when all that was available in program material was mono some loudspeaker designers went to extremes to make complicated systems.

Different parts of the frequency response would come out different parts of the loudspeaker, widely separated, giving what was advertised as a "sense of spaciousness." Maybe that was an acceptable substitute for today's stereo then. Where stereo makes the total sound associated with different instruments come from different places, this system made different frequencies come from different places.

But imagine what that did for speech. Speech does not have any really low frequencies in it. Most of the energy in speech would come from a midrange unit in one part of the system, while the intelligence part, particularly the *s* and *t* sounds, would get "squirted" in from somewhere else. Fortunately that is a bygone era now.

But understanding what it did helps us to realize the need more accurately. You cannot just throw together units to make up a multiway system and expect it to sound right automatically just because you have made sure that all the frequencies are present. There is more to it than that.

We will assume that each unit has been engineered so that it reproduces the frequencies for which it is responsible as uniformly as possible. How well this is done will vary according to the quality of the unit, but we will assume that you have chosen a unit for each part of the frequency range that is adequate for your needs in that respect.

This means that, as well as making sure the sound distribution is how you want it for that kind of room at all frequencies, and that the various units of a multiway system are organized so that the sound is well integrated, the various units are well matched for efficiency, or if they are not the way they come, that you have the means to make them so.

This requires individual level controls, particularly on the midrange and tweeter units. It also requires that each unit is capable of handling the power level your room will need to give adequate reproduction of the kind of program you will want to listen to.

The biggest demands for power are for rock music or orchestral music. Rock music sounds loud, while orchestral music making the same power demands does not sound nearly so loud, at least if it is being properly handled. Small intimate type groups, such as country or jazz, have smaller power demands.

CROSSOVERS

Now, the various units are fed through an electrical or electronic arrangement that includes something called a "crossover." We will discuss these later in the book. For now, you need to know that their purpose is to separate the frequencies fed to the different units, so the low-frequency unit gets only the low frequencies, the mid-frequency unit gets only the middle frequencies, and the high-frequency unit gets only the high frequencies.

Assuming the features we discuss later in the book have been properly cared for, we also need to make sure that each loudspeaker unit, consisting of separate parts for different parts of the frequency range, goes together to provide a balanced total reproduction that sounds like one sound.

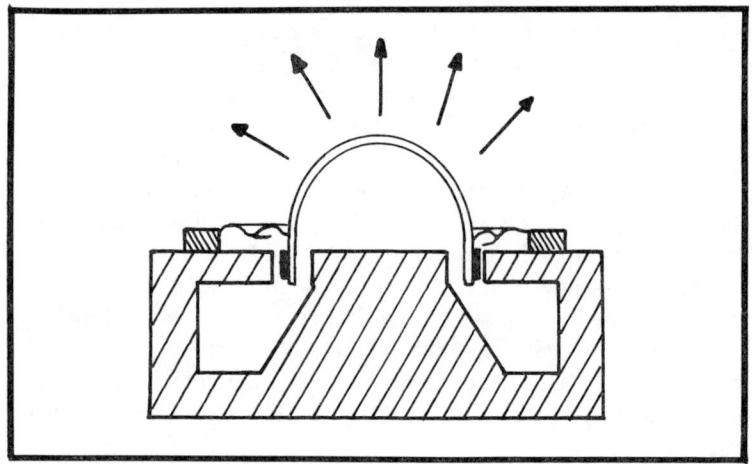

Fig. 5-6. A direct radiator with a domed diaphragm is one way to get diffused high frequencies that have a good sense of source.

The loudspeaker should not overemphasize the range coming from one of its units. If the low-frequency unit gives more than its quota, the reproduction will sound harsh or scratchy.

If the middle-frequency unit gives less than either the low- or high-frequency units, the reproduction will sound hollow, and so on. But, assuming you get all this balanced so they all give the right amount for the reproduction to sound balanced, does it come out in a way that sounds as if it all belongs together?

As we have said, low frequencies cannot be directional indoors. A foghorn at sea, where you are in wide open space, is directional. But reproduce that foghorn indoors, and its directionality is inevitably lost because of the confined space. However, all the higher frequencies, from the midrange on up, can be directional.

DIFFERENT KINDS OF DIRECTIONAL EFFECT

There are different ways by which these upper-range frequencies can be either directed or diffused, according to your needs. Uniform distribution can be achieved, but from the listening end the sounds may still be quite directional. By this we mean that they are easily pinpointed as to source. If they are fully diffused, the source is not easily pinpointed.

Some tweeters, or supertweeters, intended for handling frequencies above say 5000 hertz, use a domed diaphragm (Fig. 5-6). Its shape ensures a wide angle of radiation. But from any single listening point, you can tell quite easily where that part of the sound comes from. A high-frequency horn, which is another type used for

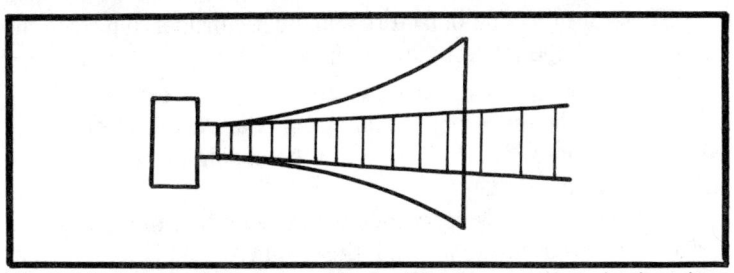

Fig. 5-7. Unless specially designed to do otherwise, most horn loudspeakers concentrate the high frquencies on their axis.

tweeters and supertweeters, will concentrate the high frequencies on its axis (Fig. 5-7).

If you sit on this axis, the high frequencies will "bore a hole in your head." Anywhere else, you do not hear them directly until they have bounced around the room a bit. Some horn type tweeters have been designed to distribute the sound better, which at least modifies this effect and may in fact remove it altogehter.

Another method of diffusing or dispersing sound is to use a device called an acoustic lens. This modifies the wavefront emerging from the unit so as to thoroughly diffuse the sound (Fig. 5-8).

As a contrast to this diffusive action, consider the multiunit directional radiator, which we will have more to say about later (Fig. 5-9). While this has the capability of concentrating the sound in a certain direction, or group of directions, from the listeners' end, the sound is diffused in a different way because it seems to come from a wide source.

In fact the diffusion from such a radiator is often so good that if you stand close to it you will think the unit is not on, and that all the sound you hear must be coming from somewhere else.

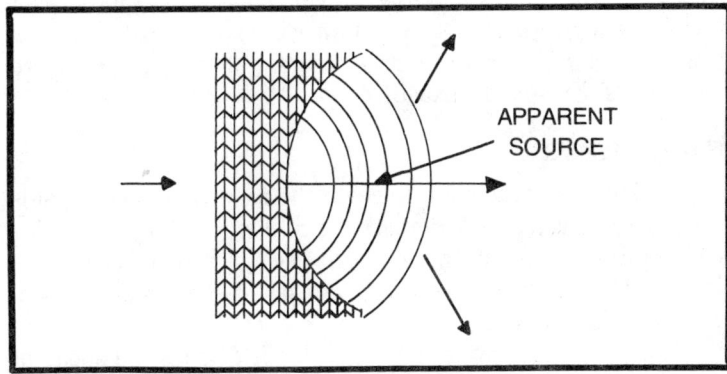

Fig. 5-8. How an acoustic lens disperses sound.

These are the kind of factors you must consider in picking the system to suit the job you have in mind.

In every installation, you have to think about directionality in two respects: at the sending end and at the receiving end. And because both are really controlled at the sending end, you have to think it through carefully.

The sending end is concerned with seeing that the total sound radiated is distributed around the room in such a way that everywhere people want to listen is properly served. But the sending end is also concerned with making sure that a suitable illusion, as to the source of the sound, is achieved at the various locations where people listen.

In some instances, the purpose of diffision is to make sure that the sound gets spread around, but retaining the loudspeaker location as the source of the sound that can be identified anywhere in the room. In other instances, such as the column loudspeaker, its purpose is to synthesize an apparent source, but usually the more important thing is to confine sound distribution to where it is wanted, eliminating unwanted reflections.

And in yet other instances, the purpose of directivity is to simulate a source that can vary according to combination with other elements and listening position in the room, such as in the system of which an example is shown at Fig. 4-15. In thinking this out, you need to be sure which you want and why, related to the kind of room and program type.

RELATIVE SIZE

Room size has tremendous effect on choice, along with furnishing, as earlier chapters have shown. Generally speaking, larger loudspeakers are better for larger rooms, smaller ones for smaller rooms. In a larger room, sounds naturally tend to get lost, or to lose their sense of intimacy. In a smaller room, you can be much more conscious of any sound's exact location.

DIFFICULT ROOMS

Do you need directivity, to cover the room with sound that is good wherever you listen? Perhaps the best way to think about this is to consider some instances where such directivity helps. We mentioned the live room often found as a family or recreation room. Why does directivity help here?

The reason we gave before was that it enables the apparent sound sources to be separated more widely in order to agree with

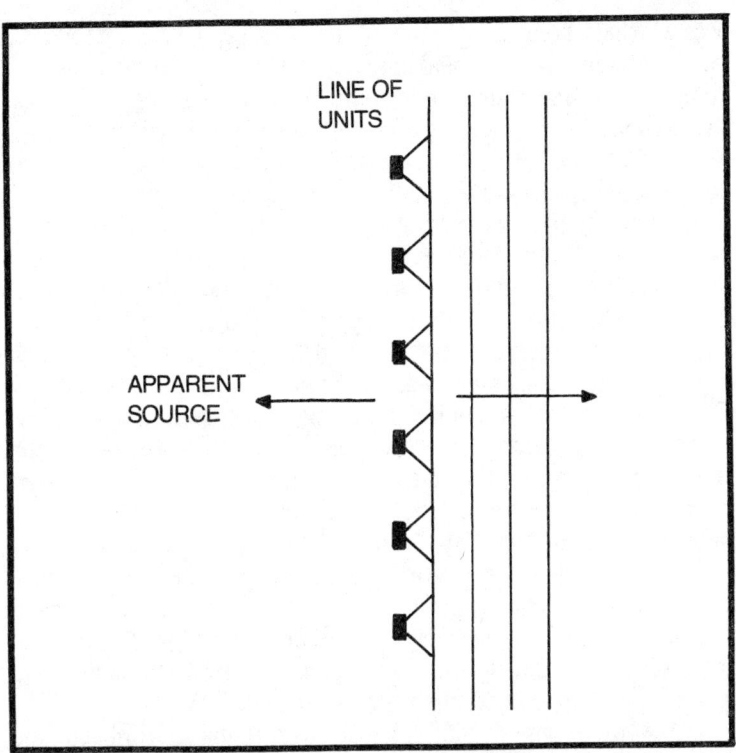

Fig. 5-9. The column, or other multi-unit radiator, beams sound in a very controlled fashion, but makes the apparent source far behind where the unit is actually placed.

how big the room seems to your ears as opposed to how big it really is. But there can be more to it than that.

Another kind of difficult room, whether of the dead or live furnishing, is one that does not have a simple shape. All the textbooks tell you where to put the loudspeakers when the room is a simple rectangle in shape. That is easy. But what about when the shape is not rectangular? A surprising number of homes have L-shaped rooms.

What you do under these circumstances depends somewhat on the proportions of the L. Think of it in terms of how a rectangular room has been modified in the simplest way to arrive at that particular L-shaped configuration. Some L-shaped rooms can viewed as a rectangular room, from which a piece has been taken out (Fig. 5-10). Others can be more easily considered as a rectangular room, to which a leg has been added.

Where the L-shaped room is formed by taking a realtively small piece out of a larger rectangle, your best approach to serving the area with loudspeakers, whether for mono, stereo, or quadraphonic, is to arrange the loudspeakers, basically as if the room was a complete rectangle, then modify either their positioning or the level that some of them work at, or both, until it sounds right in that room.

Where the L-shaped room is formed by adding a leg to a rectangle, you have a more difficult situation potentially. You may have to take separate, extra steps, to see that sound finds its way into the leg. Stereo, and even more so, quadraphonic, can get difficult in such a room. And this can become even more complicated if it is a large room with a relatively low ceiling.

One solution here, for stereo at any rate, is to alternate left and right channels around the room, so that everywhere you get a stereo illusion even though left and right change places periodically. If you have more than one left and more than one right speaker, and group them to cover areas, you cannot expect right sound to penetrate left territory, and vice versa. But by alternating them, everywhere a listener is able to hear both, which is the important thing.

You will have to use a little ingenuity to apply the same principle to quadraphonic. To figure out a way to do this, you need to think of the back channels as providing mainly ambience, while the front channels should give pinpointable sound—even if the pinpointing reverses itself in different parts of the room.

Now we turn to the kind of loudspeaker to use, according to the kind of room furnishing.

If you use diffusing type loudspeakers, sound leaves them in all directions, which includes going towards both the ceiling and the floor. In a dead room, sound that thits the floor or ceiling gets absorbed, or most of it does.

But in a live room, sound going out in these directions reflects, adding confusion to the total sound. Directive loudspeakers focus the sound toward the side walls and at the same time avoids the problem of bouncing so much off the floor and ceiling. You are putting the sound where you want it, which is where listeners will hear it.

At one time, the kind of room we will next discuss would be found only in a theater, or perhaps a closed-in grandstand, either of which few readers of this book would be installing with sound. But now that cathedral ceilings are popular in homes, that kind of room can sometimes present a similar situation. The same kind of problem can be encountered here, in a slightly different order of magnitude.

Take the worst case we ever saw: a large racetrack grandstand, where the front was enclosed with glass (Fig. 5-11). The enclosure was an enormous prism, a long volume of triangular cross section. Wherever you put the loudspeakers, it seemed the sound

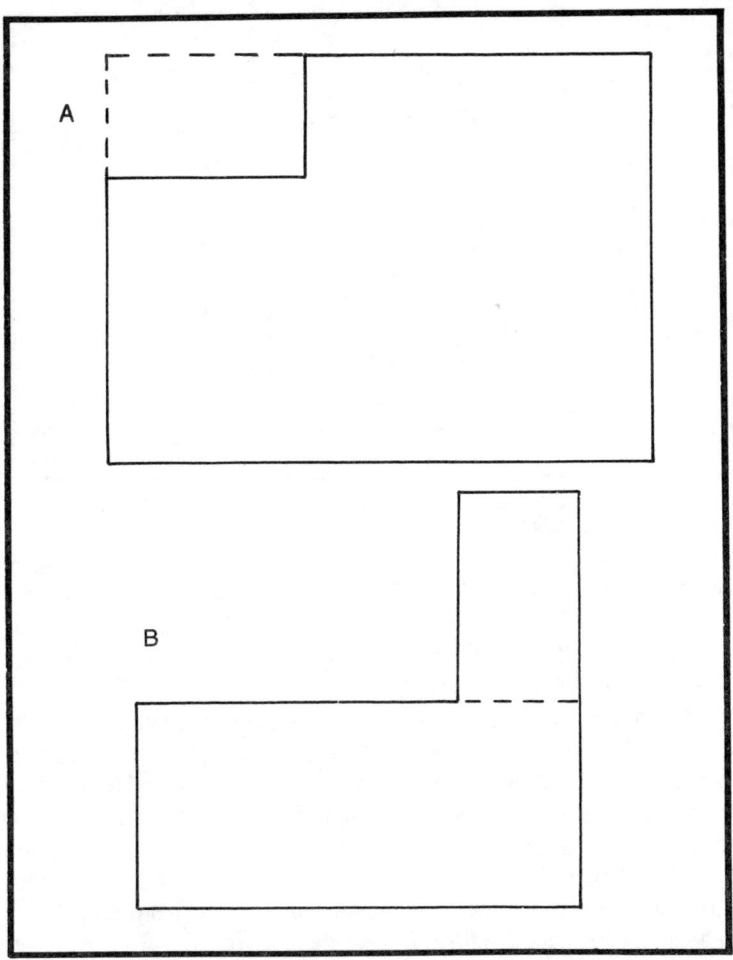

Fig. 5-10. Different ways of considering L-shaped rooms. (A) As a rectangular room from which area has been removed. (B) As a rectangular area to which a leg has been added.

bounced around impossibly so as to make any announcements completely unintelligible.

But here is the fact, use of which saved the day. When the grandstand is occupied, one side of the prism is absorptive, and that side happens to be where you want the sound, because that absorption consists of listeners. If you can direct the sound toward that side, which is the seating, without hitting either the roof or the glass front, sound should be acceptable.

What the installers did was to use directive units that beamed the sound from the apex between the roof and the glass front,

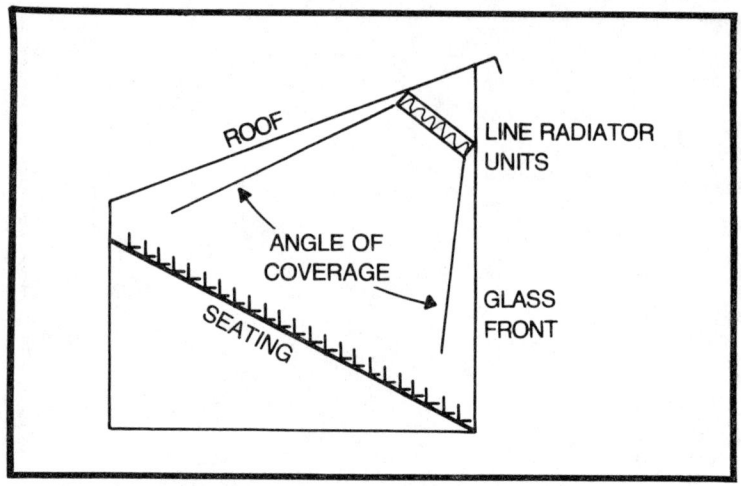

Fig. 5-11. A particular problem situation which was neatly solved by careful placement of line radiator units (columns).

directly at the seating. It worked perfectly. Sound was very intelligible when the installation was completed. The units were of the kind described in Chapter 7, although specially built for this job, so it was really quite a simple installation.

Many small theaters, or even a home where the living room has a cathedral ceiling, can pose similar problems: a large area, or volume of air, over where you listen, that can allow sound reflections to confuse the sound beyond intelligibility, or at least to make it unpleasant to listen to.

The thing to keep in mind, in thinking about how to do it, is to direct the sound where you want it, and away from where its reflection can cause confusion. That means you want to use those surfaces to guide the sound, rather than having it bounce off them. Having done this, you then seek a loudspeaker system that will provide the directivity you need from that point, or those points, to feed the listening area you have in mind.

In addition, if you want stereo, or quadraphonic, you need to have the various apparent sources where they need to be to get the proper combination of information from the various channels.

OMNIDIRECTIONALS

That usually applies to larger rooms. Smaller rooms can use a little reflection to create the aural illusion that the room is bigger than it really is. One kind of loudspeaker that is quite useful for this purpose is the so-called omnidirectional radiator (Fig. 5-12).

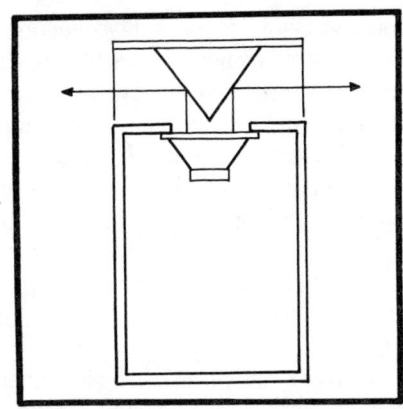

Fig. 5-12. The so-called omnidirectional loudspeaker.

All the conventional types of loudspeaker that we have discussed so far—horns, infinite baffles, reflexes, and their variations—have a front and a back, and the sound is clearer, if not louder, in front than it is at the back. And customary way to face them is toward the listening area.

But if a loudspeaker has a wide dispersion from its front, sound goes up and down, as well as out at both sides (Fig. 5-13), but not toward the back. The omnidirectional loudspeaker usually points the unit upwards. Now some of the sound goes toward the ceiling, although some use deflectors to minimize this. But sound reflected from the ceiling will create a reflection giving the illusion of a twin unit above the ceiling (Fig. 5-14).

This will reinforce sound that goes outward in all directions, and is a particularly useful system to use in small rooms with relatively

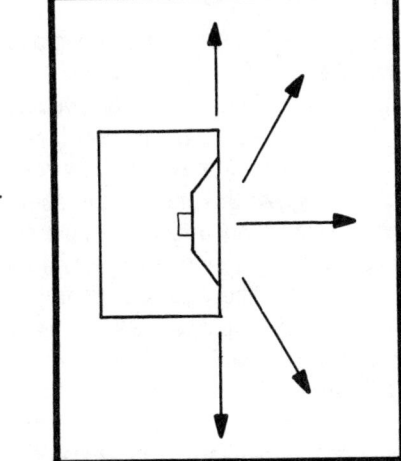

Fig. 5-13. Radiation pattern of typical closed back loudspeaker.

low ceilings. Because of the reflection pattern, the source of sound is not that well pinpointable. But it does seem to come from the right general area due to the double-image source, consisting of the unit itself and its reflection in the ceiling.

The trick to picking the right kind for the job is to have all these possibilities in your mind as you look at the room and listen to various natural sounds in it. After you have had a little practice in the proper kind of listening, you do not have to reproduce sound in the room to get the impression of what it will take to get good results. You need merely to train your consciousness or awareness to the room you are in.

As you talk, clap, or make other sounds, critical listening will tell you where sounds bounce back from. For example, if there is a bare plaster wall all along one side, by listening carefully you will be conscious of a sort of brilliance coming from the sound of your own voice from that wall.

You will notice that the brilliant wall gives you the impression that someone near you—if you listen critically, it will seem to be as far on the other side of the wall as you are on this side—is copying just your s and t sounds, those with high-frequency components in. In much the same way, as you use a handclap as a test sound, listen for how the sound of the clap gets modified.

You have probably found that, by cupping your hands in different ways, you can alter the tone of a handclap. You can make it high pitched and crisp, by holding your hands as flatly as possible. By cupping them so that, when you bring them together, you enclose more air in the space, you give the clap a deeper tone.

Have you ever tried throwing rocks into pools along a river? The plonk sound made by the rock hitting the water differs from a high-pitched plink to a very deep plonk. What makes the difference? Is it the size of the rock or what? You will find that the size of the rock affects more how loud a plonk you get, than the apparent pitch of it.

What controls the pitch is how deep the water is. The sound is due to a sound wave created by reflection from the river bed. If you can recognize the pitch of the sound, you can calculate the depth. Sound travels about 5000 feet per second in water. The plonk sound occurs at a frequency that makes the depth of the water a quarter-wavelength, like a closed organ pipe.

So if the pitch of the sound is about 100 hertz, a wavelength in water would be 50 feet, and a quarter-wavelength would be about 12½ feet, which is the depth that sound would indicate.

We mention this, because you can get similar effects in rooms, where the medium is air. If it is noticeable, this suggests that the

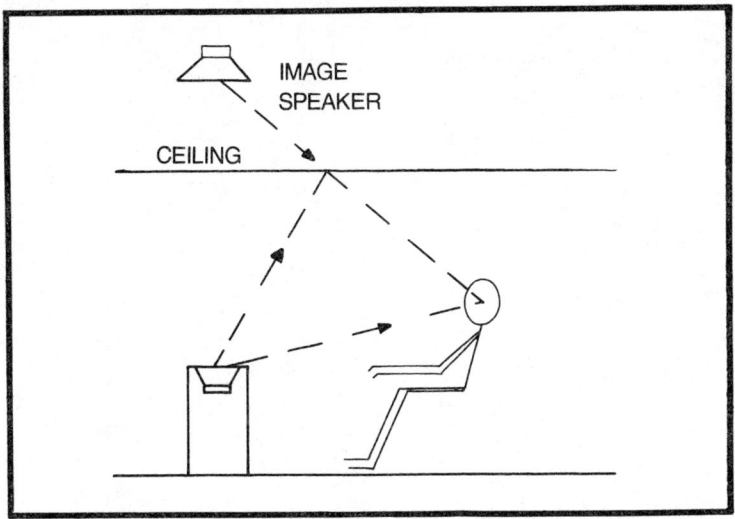

Fig. 5-14. How the image in the ceiling can sometimes help get what you want.

room could be subject to standing waves, spoiling the reproduction. The remedy for this is careful speaker use and placement so the standing waves are not stimulated so easily.

Using your hands in the room to explore its characteristics by means of handclaps is a good way to avoid making mistakes. It will take a little practice. As you become proficient, you will pay attention to all kinds of sound, even footsteps and the reflection of your own voice, coming back from the room.

This is easier than you may think, and well worth the effort to acquire. To someone who has not trained his listening faculty in this way, you may appear to have almost magic capabilities. But really it is a matter of learning to really listen for things you have been hearing all along.

6
Ways to Install Simple Inexpensive Units

You look at the low-cost unit catalog and find that they come in round or oval. Which do you want? The obvious thing is that a round unit fits a round hole, or space, and an oval unit fits an oval hole, or an elongated space, in which you could put such a hole. But that may not always be the best reason for making such a choice.

ROUND OR OVAL

If you are going to cut a hole and mount the unit flush, you are in fact putting it in a baffle. So the size or shape of the diaphragm may not be as important as you think, or its importance may take a different direction from what you are thinking. A smaller unit, when mounted in a baffle, will give more bass, or low-frequency response, than a somewhat larger unit without the benefit of such baffling.

How does the baffle make the difference? By stopping the air from shuffling round the edges of the diaphragm. Putting the unit in a baffle reduces diaphragm movement at low frequencies and increases the sound pressure it can radiate into the room. So when it is put in a baffle, its shape, and to some extent its size, is much less important for the low frequencies.

At the highest frequencies, where the diaphragm looks relatively large because it is more than half a wavelength of the frequency in question, both oval and round shapes tend to concentrate the sound on their axis. Suppose the unit is 3 inches across. Then frequencies beginning above about 2000 hertz will be beamed more

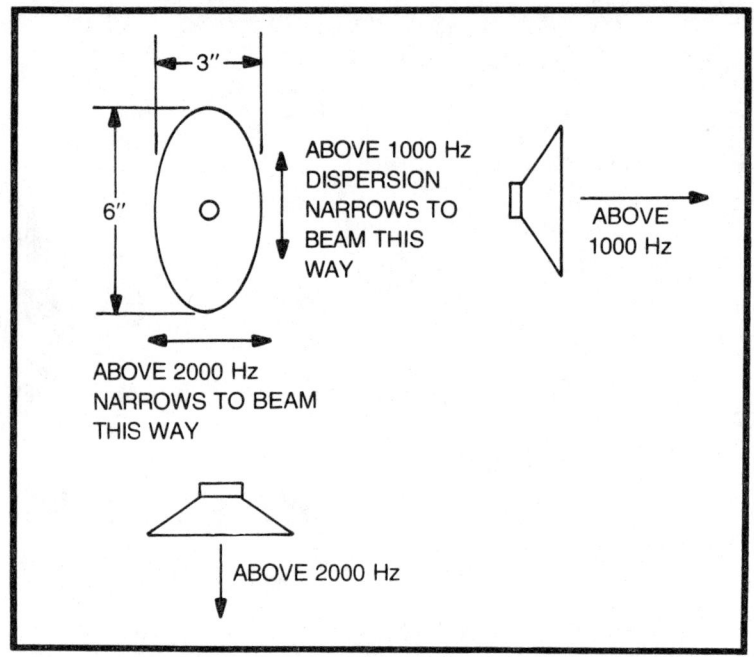

Fig. 6-1. A closer look at an oval loudspeaker unit to see what it really does to prospective performance.

and more on axis. If the unit is 6 inches across, frequencies above which sound is this beamed will start at more like 1000 hertz.

But now, suppose your unit is oval, 6 inches across the long way by 3 inches across the short way. Above 2000 hertz, sounds are beamed on axis from all directions. Above 1000 hertz and below 2000 hertz they are beamed the long ways of the unit, but still diffused in the crossways direction (Fig. 6-1).

So if you mount the unit with its long dimension vertically, in a small vacant space in some paneling, the range between 1000 and 2000 hertz will be beamed into a horizontal area, kept from straying toward floor or ceiling (Fig. 6-2). It is important to get this picture clearly in mind, if you are to use this kind of diaphragm shape intelligently.

What this means is that just about an octave of frequencies, which is really not a wide range, gets concentrated more one way than the other. And where this changeover occurs depends on the dimensions of the oval. For a 3- by 6-inch oval, the frequency range where this occurs is between 1000 and 2000 hertz. If you use a 4- by 8-inch oval, the frequency range will change to the octave between 750 and 1500 hertz.

Psychologically, you would probably expect sound to emerge in the same way the diaphragm is shaped. You might think it would spread longways and be restricted crossways. But it is just the reverse over the frequencies where anything like this happens. The sound spreads crossways, relative to the short dimension of the unit, and is restricted longways, which is the long dimension of the unit.

However, this distinction occurs only within the range where the dimensions of the unit are critical, in the example we used, between about 1000 and 2000 hertz. Below that range, distribution is not appreciably directional. Above that range, concentration tends to be entirely on axis.

Now you know what they do within that somewhat limited part of their frequency range. The "body sound" of what you want to listen to will occupy frequencies below that part of their range, although what they do in that part and above will be important to intelligibility aspects of your listening.

Because they contain the body of sound to which you listen, you are more conscious of lower frequencies. But if you remove the higher frequencies (such as by turning the tone control down) sound loses its intelligiblity and becomes muffled. So if the higher frequencies are not distributed in a way that suits your room, the sound will be good in places and bad in others.

Where the high frequencies are beamed, which will usually be on the axis of tweeter units, high frequencies will be too strong, giving a harshness to the reproduction. And when you move well away from the axis, the loss of high frequencies will make the sound very muffled.

What you are then apt to do is to adjust the tone control to suit the place where you are listening, which may be where the control

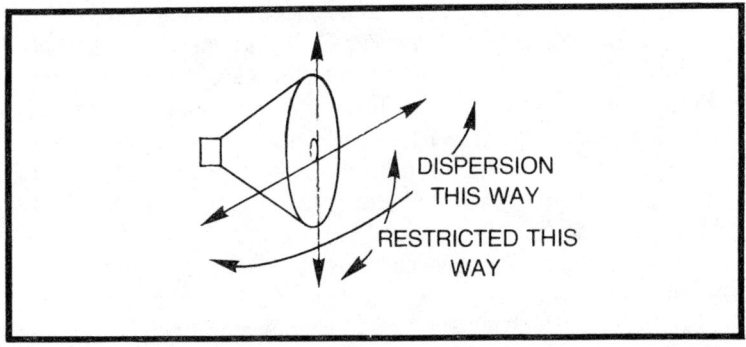

Fig. 6-2. How to visualize what an oval unit does in the limited frequency range where it does have directional capabilities that depend on its shape.

panel happens to be. This means it depends on how you face the speakers relative to that spot. If you are on axis when you adjust the tone control, you may make it sound good there, but it will be even more muffled everywhere else. And if you adjust the tone control from a position where you are off axis, a position on axis will just about bore a hole in the poor listener's head with the excessive highs.

Once you are aware of the differences to expect, you can listen intelligently to determine what you think is best. In particular, move around as you listen so you can see how sound varies in different parts of the room.

The important thing is to get enjoyable sound where you want to listen to it. Try to get it so that any normal listening position gives you good sound. If that is not possible, try for the best compromise.

GETTING THE BASS

In all circumstances, you must also consider how you are going to get the lower frequencies out into the room. Unless you are ambitious enough to construct some kind of horn, you will more likely do this by using some kind of baffle effect. But there are other effects you can sometimes use.

People who design loudspeakers think in terms of a diaphragm stiffness and mass, the efficiency of the transducer (that is the moving-coil mechanism that drives the diaphragm), and the acoustic properties of the box, baffle, or whatever the unit is mounted in.

You may read articles by professional designers saying how they made the best possible loudspeaker using a certain design combination of one kind or the other. Most professional designers have all the variables well in mind. There are too many of them for most of us to keep in mind, although we can try.

The result is that many people read articles like that, then try to form their own conclusions following the same approach. Starting from the basic parts and putting it all together is apt to lead you to the conclusion that the way you just figured out is super. You cannot easily compare it with other kinds of design that way.

If you work your way through each way, and come to a super design, how do you really tell which is best, or best for your particular situation? It is difficult.

Really, we find it easier—and we think you will too—to think backwards with respect to normal procedures. Of course, how efficient the unit is, will affect how well that part of the system does its job. But what we want to discuss are the basic limits related to the outside of the unit. It has to move so much air to make the various

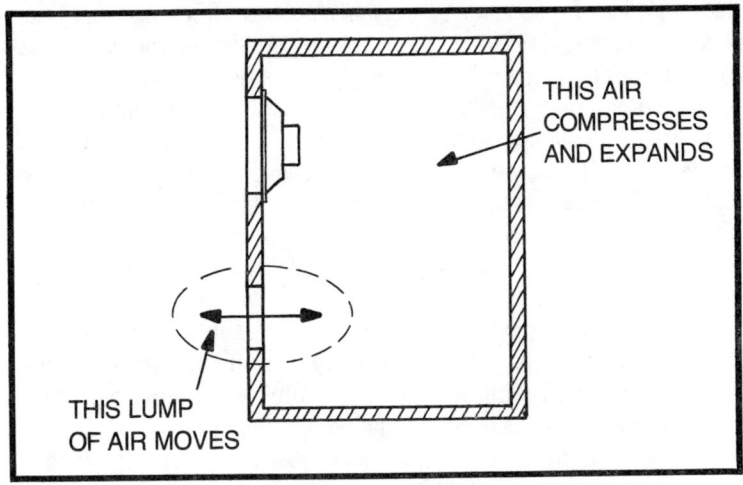

Fig. 6-3. How to think about the air in the various parts of a bass reflex enclosure.

sized waves, from the biggest for the lowest frequencies on down to the smallest for the high frequencies. There just is no way around that.

At the lowest frequencies, where the room itself that you are in is smaller than one whole wave of the frequency you are thinking about, starting a wave across the room is not the object. You have to produce the necessary sound pressure fluctuations inside the room.

But as you go up in frequency, you will always come to frequencies where several waves are propagated into the room. When you reach these frequencies, the size of the box becomes a factor, whatever you may have inside it. For this reason it is a fairly safe rule that the size of the loudspeaker should be somewhat in proportion to the size of the room in which you plan to listen to it.

This does not mean that you can write a formula to say that the box should have a volume that is some definite fraction of the room's volume. For example, a speaker in the shape of a 1 foot cube would have one thousandth the volume of a room that is a 10-foot cube. The relationship between the sizes will vary with size too.

About the only thing you can really say is that a larger room will need a loudspeaker with a larger box, and vice versa. This is the thing to remember.

THE BOTTOM END

The biggest problem is always with the lowest frequency you want to reproduce. There is one more ingredient that you can use:

the same kind of resonance that organ pipes use. Isn't this what bass reflex enclosures do? No, it isn't, and we need to see the difference to avoid making some mistakes on this question.

An organ pipe resonance is determined by the length of the pipe, not its volume. Resonance occurs in a stopped pipe at a frequency such that the pipe is either a quarter-wavelength long or an odd number of quarter-wavelengths long. Thus a 10-foot stopped pipe will resonante to a frequency with a wavelength of 40 feet, which is about 27 hertz. It will also resonate at odd multiples of this, such as 81 hertz, 135 hertz, 189 hertz, and so on.

An open organ pipe, that is one where the top is open as well as the voiced end, produces resonance for a frequency such that the pipe is a half-wavelength long, or a multiple number of half-wavelengths. Thus a 10-foot open pipe will resonate to a frequency with a wavelength of 20 feet, which is about 54 hertz. It will also resonate at all multiples of this, such as 108 hertz, 162 hertz, 216 hertz, and so on.

Notice that the frequencies to which stopped and open pipes of the same length resonate sort of interlace. The important thing, relative to comparing the resonance of a bass reflex cabinet with an organ pipe resonance, is that for all forms of organ pipe resonance length is the important dimension because resonance is related to propagation velocity of sound. In a bass reflex cabinet, volume and opening size are important because that behaves like a Helmholtz resonator.

The Helmholtz resonator's frequency is not related to propagation velocity. And also, unless you somehow make it a hybrid so that it sometimes behaves like an organ pipe, the Helmholtz resonator has only one resonant frequency, where organ pipes have a whole succession of them. This gives them their various richness in overtone structure that a bottle resonator does not have.

So there is a fundamental difference between the two kinds of resonator. To explain this, we need to think how the air behaves inside of them.

The ordinary bass reflex unit is small enough, physically, that you think of the air inside the box acts like a spring. The movable air in the port acts like a weight. What you hear is due to the movable air.

All the while the box is relatively compact in its shape, the important dimension controlling the compressibility of the air inside the box it its volume. You can take a design that you know works with inside dimensions of, say 20 by 24 by 12 inches, which multiplies out to 5760 cubic inches, or 3.33 cubic feet. Now you can change the shape if you want, keeping the volume at 3.33 cubic feet, without materially changing the performance.

For example, if the 12-inch dimension protrudes into the room too far, you could change that dimension to 8 inches and make the other dimensions, say 24 by 30 inches, which again multiplies out to 3.33 cubic feet. Or, if the 20-inch dimension is too wide, you could increase the depth to 15 inches, making the other dimensions 16 by 24 inches, which again multiplies out to 3.33 cubic feet.

One thing to note about picking dimensions is that best results are obtained when all the dimensions are a little different. In other words, a square, and even more particularly a cube, is not a good shape for a loudspeaker enclosure because that makes for a very definite resonance, or coloration of the sound, at the frequency for which that dimension happens to be a quarter-wave, a half-wave, or some other multiple.

We mentioned just now that, under some circumstances, a bottle or box resonator can behave more like an organ pipe resonator, in that a single dimension can be important, rather than the volume. This can always happen along any dimensional direction of the box.

If you make two or more dimensions the same, this will produce extra emphasis at whatever frequencies that dimension favors with resonances.

Thus, if you make the box 24 inches each way, the resonant wavelength will be 4 feet, or 8 feet, which corresponds to a frequency of 1100, divided by either 4 or 8, which is 275 hertz or 137.5 hertz. Having one dimension more and another less to get the required area or volume, so that all three dimensions are different, spreads any coloration to the point where it is not noticeable.

BUILT-IN RESONATORS

In the struggle about size, there are some tricks you can use to cheat a little bit on the principles we have discussed so far. Here is one that you can use for built-ins. If you mount the speaker unit in the wall, or ceiling, you can use the spacing between studs, or joists, as a sort of organ pipe (Fig. 6-4). Now the important dimension is not the width between the members, or the volume of the enclosed space, but the length of the space you use.

Now what you are in effect doing is putting the loudspeaker unit in the mouth, or voicing end, of something that closely resembles a stopped or closed organ pipe. That is why length becomes important now, rather than volume.

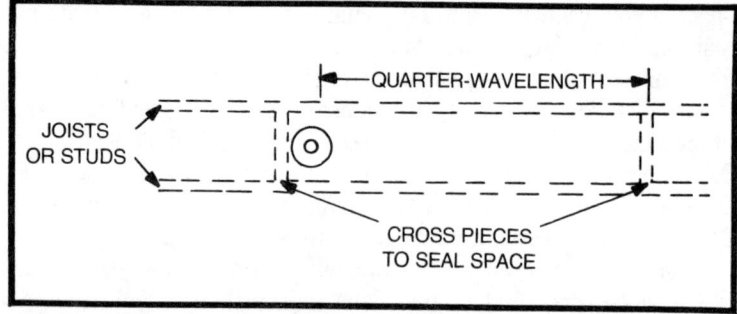

Fig. 6-4. The idea of using spaces between wall studs or ceiling joists to make elongated enclosures that function quite differently from either infinite baffles or reflexes, using a quarter-wave resonator principle.

Suppose the space is 8 feet long. In this configuration, if you have an opening at one end of the 8 feet, the resonance will be at a frequency that makes 8 feet a quarter of a wavelength. Thus the full wavelength will be 32 feet, which represents a frequency of 1100 divided by 32, or about 34 hertz.

If the loudspeaker unit has a diaphragm with a resonance well above 34 hertz, in any conventional box its response would be virtually dead at a frequency as low as 34 hertz. But a broadly tuned resonator (because your opening will not be so sharply defined as that in an organ pipe) can pull up the response in that region quite well.

But the response may have a dip between the unit's resonance and the organ pipe resonance (Fig. 6-5). To overcome this, if you have the space between ceiling joists to use and have say 14 feet length that you can use, you could choose a spot a little off center, so you have two pipes, one extending say 6 feet and the other 8 feet for the hole (Fig. 6-6).

This would reinforce frequencies at about 34 hertz for the 8 feet section and 45 hertz for the 6 feet section, somewhat broadly, to give a better bass response to an otherwise not-so-good unit for bass.

The point to realize, in figuring this kind of thing out, is that everything is relative. When you think of free air out in the room, the diaphragm is stiff and restricts air movement below its resonance, which may be at 90 or 100 hertz. Normally, the response will die below that resonance. But when you put the same unit at the mouth of a resonant tube, formed by the space between two joists or two studs, the stiffness of the diaphragm no longer compares to that of the free air in the room. It is working against the mass of air in the hole at the end of one or two pipes.

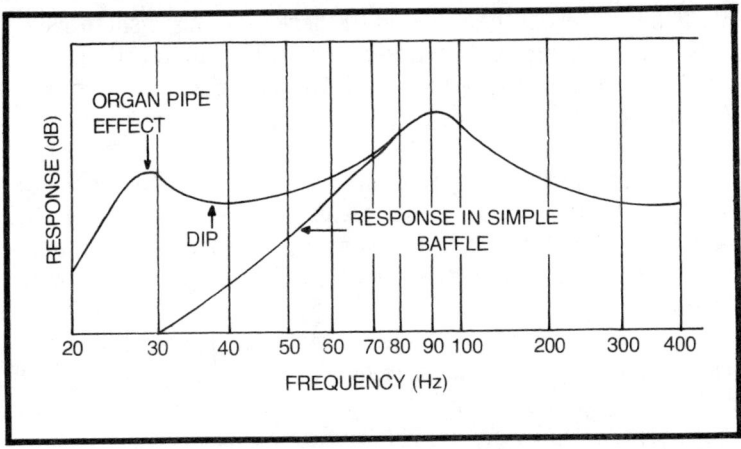

Fig. 6-5. How the quarter-wave resonator can augment the low-frequency where it counts.

The air here moves more readily at the pipe resonance, so the diaphragm mass now seems smaller in comparison with the mass of air that moves with it at this frequency than it would in a box of another shape. So you get a boost in response at that frequency, or those frequencies, where you use a double pipe.

The frequency we have been calculating has been based on the concept of a stopped organ pipe. In a stopped organ pipe, it has two ends, one stopped, the other open. The voice end communicates directly with outside air. In fact, that is where the sound comes out.

An open pipe, similarly has both ends open. But when you put a loudspeaker in one end, strictly speaking it is neither open nor closed completely at that end. Because the loudspeaker cone moves fairly freely, it comes much nearer to looking like an open end than a closed end, thus the calculations based on the organ pipe will not be far off.

But the presence of the loudspeaker cone will make a difference to the resonance frequencies of these improvised pipes.

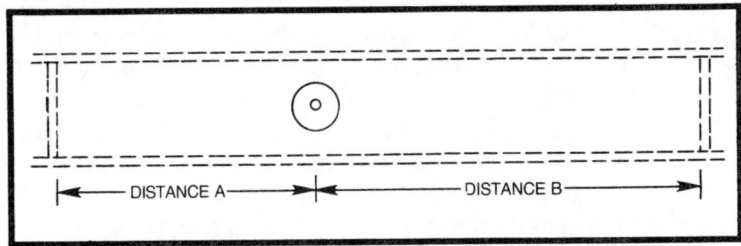

Fig. 6-6. If the space available is long enough, you can utilize a double quarter-wave resonator effect with even better results.

If the unit resonance is around 100 hertz, you may want to make the shorter, second pipe split the difference a little better. Perhaps it would be more appropriate to divide a 15-foot length into 5 feet and 10 feet, putting resonances at 27.5 hertz and 55 hertz, approximately.

The natural diaphragm resonance will be somewhat indefinite when mounted in this kind of a box because not all of the air behind it is coupled to the diaphragm, as you approach what would be resonance, if it were mounted in a more compactly shaped box of similar volume.

Then again, the fact that wave travel, back and forth along the pipe, has put diaphragm resonance above the resonance of the pipe, means that you should think about secondary resonance effects. These occur at odd harmonics of the first resonances.

Thus, using the 15-foot space, divided into 5- and 10-foot sections, the 10-foot section would resonate at about 82.5 hertz, as well as 27.5 hertz, and the 5-foot section would resonate at about 165 hertz as well as 55 hertz. So you have a fairly broad sequence of resonances at 27.5, 55, 82.5 and 165 hertz.

The properties of the loudspeaker diaphragm will shift these theoretical frequencies a little, but these calculated values make good starting points from which to experiment.

If you have access to the space between the joists, while you are doing your built-in, you can try blocking off sections of joist space in opposite directions, at various distances, to see how it sounds, and adjust them for the smoothest sounding low end.

You will have to fasten the crosspiece that stops the pipe in place, and make sure that it seals the space at that point, then replace the flooring or wall paneling, or however you got access to the space, to give it a tryout. Just like the bass reflex, a pipe should be properly sealed at all points.

So to make adjustments, you need to make a way to quickly get into it and secure it again for each tryout. If you have a gliding tone frequency test record, so much the better. If not, use some recorded music, that has a good variety of different bass notes in it. Watch for the adjustment that gives you the most uniform response to all the bass notes, without overemphasizing some of them.

That applies the principle of the closed, or stopped organ pipe, for loudspeaker work. Can we also apply the open pipe principle? There may be a way, but we should comment about a system we once observed demonstrated, for which the man demonstrating it had a patent. As this was 30 or 40 years ago now, the patent must have expired for whatever it may have been worth.

The device looked like a small pipe organ: a bunch of pipes, all of different lengths, with their top ends open. Their bottom ends were also open, but snugly mounted in the top of a chest, in which was mounted the conventional loudspeaker unit. Thus the loudspeaker unit communicated the sound waves it generated to the outside world via a wide selection of open pipes of different lengths.

The inventor's idea sounded good. In those days, the big problem with most loudspeakers was a whole family of resonant frequencies. It was known by then that a well-designed horn could do a lot toward eliminating their effect, but he wanted to use a more direct approach.

Instead of getting rid of resonances, he would use them, and just make sure there were enough so that all frequencies were treated fairly: sort of a "democratic frequency response." Well, can you imagine how it sounded.

To demonstrate it, he used a recording of organ music. It sounded terrific. The best organ reproduction we had ever heard. But then we wondered what it would sound like on speech. So we put on a voice recording. Can you guess what that sounded like? Just what you would expect if you think about it: like someone talking inside the voice chest of an organ, with his voice getting out through all the pipes.

It was virtually unintelligible on speech. So, if you want a loudspeaker that will sound good on organ music, and precious little else...you know how to go about it.

HOME-BUILT HORNS

At one time, the building of front and back loaded horns was a big thing in the home construction of loudspeakers. In the back loaded type, the upper frequencies radiate directly from the front of the diaphragm, while the back feeds into a folded horn that extends down to a fairly low frequency, lower than the unit would otherwise handle.

This is the important advantage of a well-designed horn: that it loads the cone, or diaphragm, making it transfer more of its energy into an acoustic wave, which means it is more efficient, and at the same time virtually eliminating, not only the main resonance of the loudspeaker, but also all of the lesser resonance effects that give a loudspeaker coloration.

In the old days, before loudspeakers were really designed by engineers, and before anyone really thought about taking frequency responses, you would see loudspeakers advertised as having *cathedral tone*, or *concert hall presence*, or some such catch phrase. What this was intended to convey, was that, if you let your imagination wander, you could imagine you were listening to a performance in one of those settings.

The effect was due to the set of resonant frequencies peculiar to that particular make of loudspeaker. As loudspeaker design has improved, the effort has been to make a toneless loudspeaker: one that will reproduce whatever program is fed into it, without introducing any tone of its own. The horn was the first way this was achieved and it remains one of the best ways.

One engineer explained it this way: if something has a high efficiency at all frequencies, then its frequency response has got to be good. A unit that has a low general efficiency, but with improved efficiency where there are resonances, will not have such a good frequency response. And that is true.

What happens to a loudspeaker's resonance when it is put into a horn? We have discussed what happens to other kinds of loudspeaker. When you put a unit into a horn, you impose a heavier load on its diaphragm movement at all frequencies, including resonance, so that most of the energy delivered to the voice coil is spent pushing an acoustic wave down the horn.

A well-designed horn must include attention to the part where the unit matches to the throat of the horn. A large cavity here, for example, can have its own resonance, the thing we were trying to avoid. Or we can use that cavity as part of an acoustic filter to provide a cutoff frequency, where the horn leaves off, to have a direct radiator take over.

The back-loaded horn uses waves from the back of the cone to feed into the horn for low-frequency reproduction. Then the cavity provides an acoustic cutoff, like a low-pass filter, allowing the front of the cone to radiate frequencies from this point on up.

Below this acoustic crossover, the horn loading on the back prevents any appreciable cone movement, so the front seems dead to these frequencies. Above the acoustic crossover, the cavity absorbs energy at the back, so it no longer goes down the horn. This permits the cone to move more freely to allow radiation from the front.

But whether back loading or front loading is used, these designs were very carefully engineered to provide the desired overall performance. You could buy them ready-made, or you could buy plans and make one yourself.

These units were usually built to a design purchased for the purpose, or copied from some similar design. While making a perfect horn involves quite critical calculations and manipulation of the design, most practical designs that become commercial loudspeakers were approximations that performed well enough that you could not tell the difference.

HORN DESIGN

While the technical basis for the design of an exponential horn is quite mathematical, it can be simplified to some easy-to-work figures. The lowest frequency any horn will handle, called its cutoff frequency, depends on two things, both of which must be satisfied for the horn to work right down to that frequency. The first is the size of the mouth. The second is the rate at which it flares.

Let us illustrate this with a few examples. A simplified formula for mouth size requires the diameter, across the mouth in each direction (which can be either round or square), to be a dimension obtained by dividing 4000 inches by the cutoff frequency. Thus for a horn with a 100 hertz cutoff, you need a mouth 40 inches across. For a 200-hertz cutoff, this comes down to 20 inches. For 50 hertz, it goes up to 80 inches.

You see why a corner horn needs the walls and floor (or ceiling) to make it work right (Fig. 5-4): only out in the room does the mouth reach 80 inches across, assuming it has a 50 hertz cutoff. If the cutoff frequency is lower, the minimum equivalent mouth size is correspondingly larger.

The flare rate can most easily be calculated by taking the distance along the horn, from throat toward the mouth where the sound wave travels, at which the area regularly doubles. The formula for this distance can be taken as 700 divided by cutoff frequency. Thus the 100-hertz cutoff horn should double its area every 7 inches. For a 200-hertz cutoff this comes down to 3½ inches, while for a 50-hertz cutoff it goes up to 14 inches.

Now let us see what this means if we know how big the throat has to be. It should be about one-third the area of the diaphragm that it couples to. So if you plan to use an 8-inch speaker, whose cone area area will be about 50 square inches, the throat area should be about 16 squre inches, which would be a 4-inch square throat.

Now, if you want a 100-hertz horn, the mouth must be 40 inches across, making an area of 1600 square inches. The mouth area must thus be 100 times the throat area. It must double every 7 inches, so count up how many 7-inch lengths you need: in the first 7 inches it grows to 32 square inches, in the second 7 inches to 64 square inches, in the third to 128, in the fourth to 256, in the fifth to 512, and in the sixth to 1024. The seventh would go to 2048 square inches, but we only need 1600, so about another 5 inches will be enough, making the total length of the horn 6 × 7 + 5 = 47 inches.

The flare rate can most easily be calculated by taking the distance along the horn, from throat toward the mouth where the sound wave travels, at which the area regularly doubles. The formula for this distance can be taken as 700 divided by cutoff frequency. Thus the 100-hertz cutoff horn should double its area every 7 inches. For a 200-hertz cutoff this comes down to 3½ inches, while for a 50-hertz cut-off it goes up to 14 inches.

Now suppose you will accept a 200-hertz horn. Starting with the same 16-square-inch throat, you now need to get up to only 400 square inches (4000 inches divided by 200 equals 20), and it doubles every 3½ inches, which will go 32, 64, 128, 256—that takes 14 inches, and another 2 or 2½ inches will get you to 400 square inches. So for a 200-hertz cutoff, the horn length comes down to 16 or 16½ inches.

But to get down to 50 hertz response, you need a mouth area of 80 inches square, or 6,400 square inches. And it now takes 14 inches to double area each time (700 divided by cutoff). So you start with 16 square inches again and, at 14-inch intervals, the area should be 32, 64, 128, 256, 512, 1024, 2048, 4096; that is eight intervals of 14 inches. The next one would go to 8192 square inches, but we need 6400, so we need 9 or 10 inches of that interval, a total length of 8 × 14 + 10 = 122 inches—more than 10 feet.

These three possible designs are tabulated and illustrated for better comparison, using a square section horn all the way, in each case, in Table 6-1 and Fig. 6-7.

The 100-hertz horn needs to be almost 4 feet long. The 50-hertz horn needs to be just over 10 feet long, while the 200-hertz horn is not much over 1 foot long. We have shown what this means

Table 6-1. Comparison of Three Horn Designs.

100 Hz Horn			200 Hz Horn			50 Hz Horn		
Distance from throat (in.)	Area (sq. in.)	Side (in.)	Distance from throat (in.)	Area (sq. in.)	Side (in.)	Distance from throat (in.)	Area (sq. in.)	Side (in.)
0	16	4.0	0	16	4.0	0	16	4.0
7	32	5.7	3.5	32	5.7	14	32	5.7
14	64	8.0	7.0	64	8.0	28	64	8.0
21	128	11.3	10.5	128	11.3	42	128	11.3
28	256	16.0	14.0	256	14.0	56	256	16.0
35	512	22.6	16.5	400	20.0	70	512	22.6
42	1024	32.0				84	1024	32.0
47	1600	40.0				98	2048	45.2
						112	4096	64.0
						122	6400	80.0

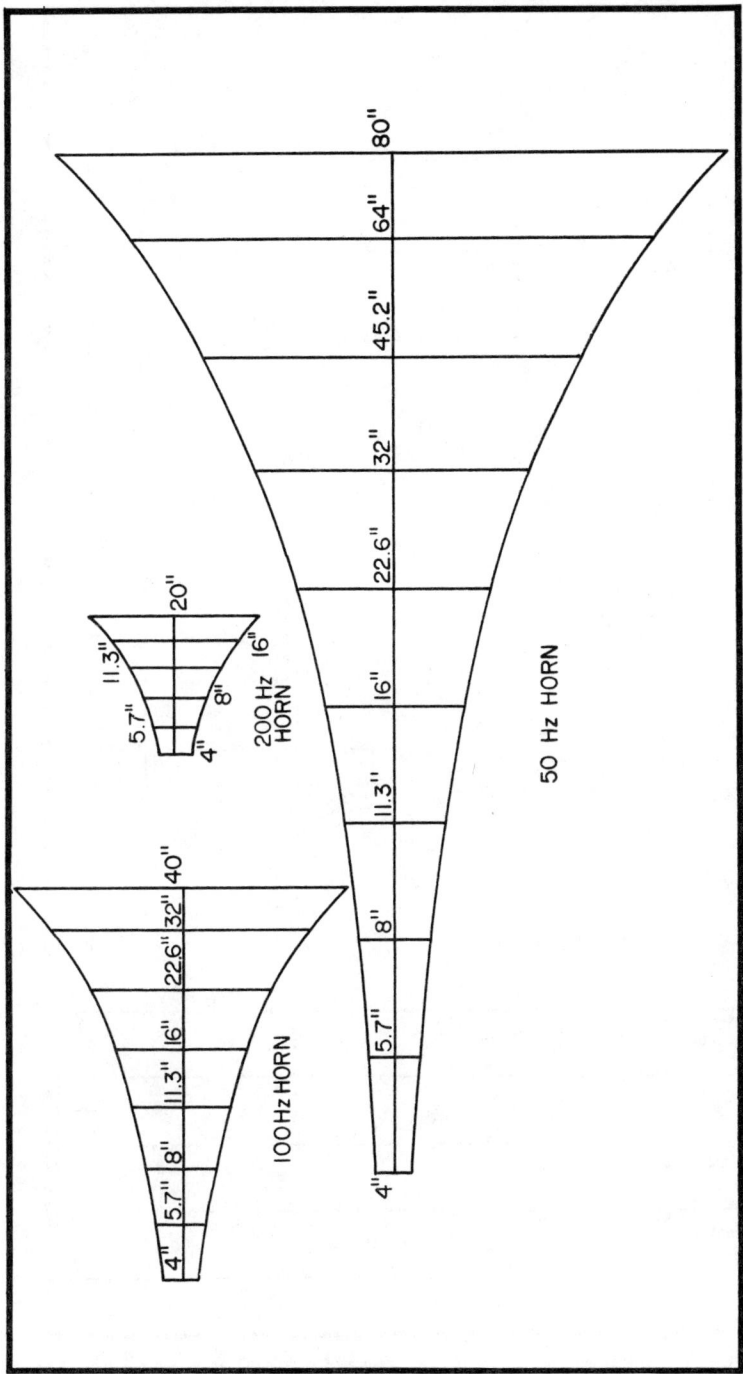

Fig. 6-7. Comparison, to scale, of a 100-hertz horn, a 200-hertz horn, and a 50-hertz horn, all designed to use the same unit at its throat.

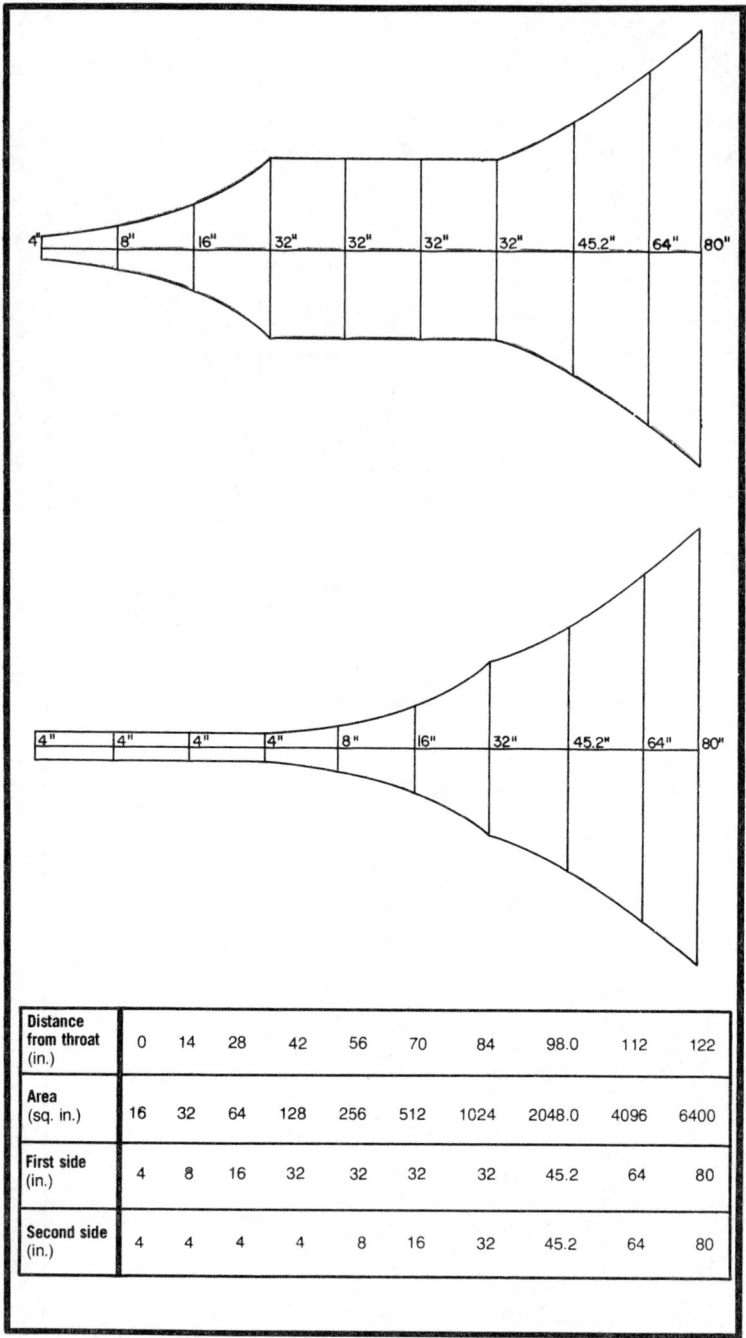

Fig. 6-8. How to lay out sections in the design of a folded horn.

along a straight horn. To make this into a corner cabinet, you need to find a way to fold the horn so it doubles back on itself.

In the expansion, you do not have to keep the section square so long as the area where it delivers the wave at its mouth has the minimum dimension specified in each direction. But in the expansion, the section can stretch first one way, then the other. The area expansion should be smooth.

You can use sloping sections and calculate the area at each distance at which it is supposed to double (Fig. 6-8). Avoid any sudden steps in the shaping as measured by area. When you go round corners to double back, try to figure the average distance the sound wave will travel in negotiating that turn (Fig. 6-9).

On the inside of the turn, that part of the wave will "mark time," while the part that goes round the outside travels double the average distance. Because the air is contained in an expanding tube, provided the frequencies you are considering are such that wavelengths are bigger than the difference between inside and outside path length, the air behaves like a fluid, averaging the flow rate of the wave around these bends.

Of course, the air itself does not actually flow; it is the sound wave that moves. But it moves by vibration of air particles, back and forth along the direction of flow, which is just like the air itself flowing, providing the wave dimensions are large, compared with the horn dimensions at that point.

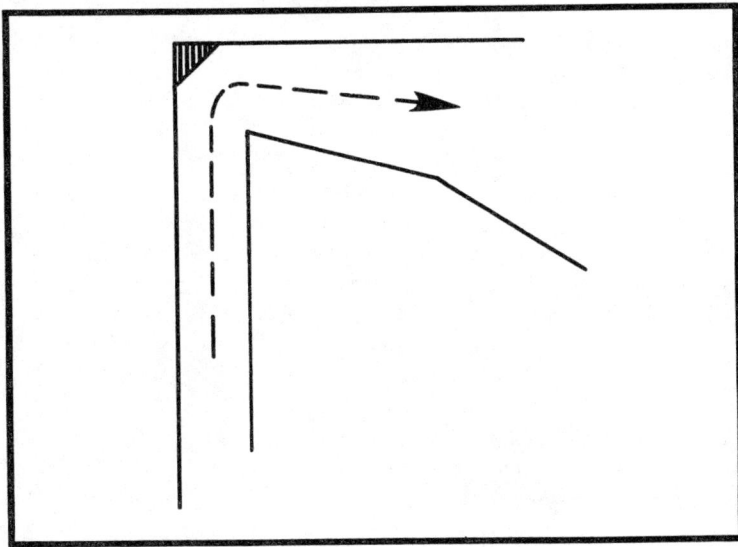

Fig. 6-9. How to calculate distance along the horn, in negotiating a turn or fold.

Actually, to be precise the distance unit of measurement for a particular frequency is the wavelength at that frequency, divided by 2π. So for 100 hertz, the unit of distance is about 11 feet, which is the wavelength, divided by 6.28, or about 1.75 feet, which is 21 inches. You are looking at area doubling every 7 inches (700 divided by 100), which is a fraction of 21 inches.

7

Ways to Install Multiple Simple Units

One of the great things about mass production is that making large quantities of something brings the price down. This is a principle that we can use to great advantage in this particular way of designing loudspeaker coverage. To see why it works so well, let us take another look at the problems we face.

The one problem that has plagued loudspeaker designers most is that of handling the enormous range of audible frequencies, with their associated wavelengths, a range of almost a thousand to one. For the low frequencies, we need a big unit to push all that air for the big waves. For the high frequencies we need a small one to do each frequency justice without discriminating in order not to make the response ragged up there.

The big units do not handle the high frequencies well, and the small units cannot handle the low frequencies at all—not by themselves, anyway.

ONE WAY OF GETTING IT TOGETHER

This is one of those solutions that seems ridiculously simple once someone has told you what it is. It solves both the cost and the frequency range problem, and avoids the need for a crossover, which any multiway system needs. It is simply to put a number of small units all on one baffle board, or the equivalent part of an infinite baffle.

Suppose you mount 37 loudspeaker units with 6-inch diaphragms in a hexagonal pattern (Fig. 7-1). The whole thing will be more than 42 inches across in every direction. Now think what you have when all these are connected so all their diaphragms drive together. We will see how to do this in Chapter 11.

At the higher frequencies, each unit handles itself as well in combination as it does alone. And for the lower frequencies, the combination, all working together, will push as much air as a single unit with a diameter of 42 inches. How can that be; won't they still have the same resonant frequency, thus the same low-frequency cutoff?

As we have pointed out, a loudspeaker's resonance is affected by the load the air places on the diaphragm, as well as just the weight of the diaphragm itself. If you put one of these small units in a baffle by itself, you lower its resonance because you are making it move more air. When it works by itself, it moves air all over the surface of the baffle.

When you put it in between other units, all of which are driven together, they move the air in front of them, and the unit you are thinking about moves air only in front of it. So to get a corresponding movement, it must move air further out from itself when the wavelengths are long. This is what pushes the resonance frequency of the whole system down a lot further, and it behaves like a much bigger unit.

Look at it another way. Compare the total mass and total compliance or springiness of one big cone with that of all these little cones combined. The resonant frequency of each individual small speaker is at a considerably higher value, because (a) the mass, or weight of its cone is much smaller, and (b) the suspension is relatively much stiffer.

Note that word relatively. Assuming it would fit, how would the small cone suspension perform if used on the single large one? It would be quite inadequate, would it not? The first appreciable movement of the large cone would probably rip it to shreds.

Now, put together all the 37 small cones, and the one large cone: which do you suppose would weight the most? Not too much difference, perhaps, but probably the one large cone. And how do you combine all those suspensions into one compliance?

Each suspension works on its own piece of the total mass, its own cone. So the total control on the whole moving set of diaphragms is no more than the control on each individual one. It will average out, if they are not all identical. What is different from the cones working individually, is what we have already mentioned, the fact that the air in contact with them provides a sort of coupling between them.

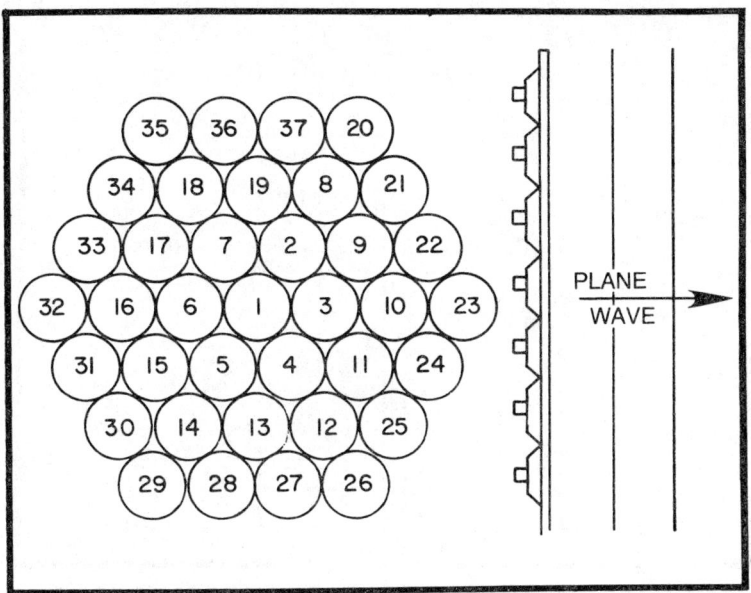

Fig. 7-1. A loudspeaker assembly that will handle a wide frequency range, and at the same time beam the sound.

If only one cone was driven, the air pressure it creates would try to drive the adjoining cones in the reverse direction to equalize the pressure. But the fact that all of them are driven, means that each takes a part of the air load, as we have already said, in front of it. Only around the edges of the hexagon does this change, because there the cones do not have further cones beyond them. This means they will move rather more air, allowing them a little greater movement than cones nearer the middle.

But the net effect, with all of them, is that the resonance when they act together, is not very different from that of the single large cone—a little higher in frequency, maybe, but not all that much. Another thing to think about is how much the diaphragm or diaphragms move. Because the big cone is so big, and its suspension allows a bigger movement than do the smaller suspensions of the small cone, we tend to think of it as using a big movement.

But in practice, this does not happen, unless the loudspeaker is radiating a very large acoustic power. Because it is so large, we are talking about something 3½ feet in diameter, it can move a lot of air with quite small movement. However, if you want to simulate an earthquake, you are talking about a big movement, which the big unit could handle and where the 37 small ones could not.

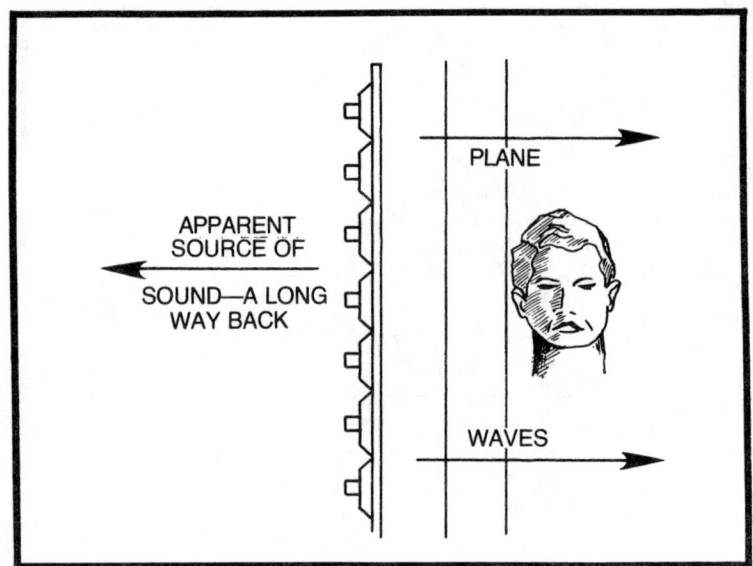

Fig. 7-2. Effect of multiple-unit loudspeaker on the apparent source of sound.

For normal bass reproduction, the 37 small units can handle the movement as well as the big one can. Once you have this understood, you see where each variety has its own advantages. The big unit is only, at best, a sort of superwoofer. If your problem is to provide earthquake effects for the movie of that name, it will do the job fine, but it will not reproduce all the rest of the sound, the higher frequencies, at all.

On the other hand, the 37 small units will handle full range, which the single big one will not.

Now, we have touched on another advantage of this method: the fact that they all move together, thus produce what is called a "plane wave." As we have already mentioned, if you stand close in front of such an array, you experience a somewhat unexpected illusion. Standing close to any single loudspeaker unit when it is working, you can hear sound coming from it. But when you stand close to one of these arrays, they sound as if they are not working.

The sound seems to come from much further back, behind them (Fig. 7-2). This is because your hearing faculty receives the plane wave radiated by all of them in unison, not a lot of separate waves radiated individually. What, in your everyday experience, would normally produce such a wave? The nearest thing would be a single unit, much further back, wouldn't it? So that is what you think you hear.

Of course, as you move away from a position directly in front of such an array, what you hear will change. Throughout the rest of the room, the sound is directed in that plane wave, which makes it travel in just one direction: forward. The spread at the sides is much reduced because the individual diaphragm movement is much smaller than a single unit would have working by itself.

So putting such a group of loudspeaker units in a cluster, facing a seating area, will concentrate most of the sound energy in that seating area, avoiding reflections from surfaces in the rest of the room or from surrounding areas (Fig. 7-3).

Such a bank of loudspeakers may prove to be less costly than a single big unit with the necessary smaller units to handle the higher frequencies, but it still shares one thing with the big unit: it is big and cumbersome.

COLUMNS

A slim line of loudspeaker units, built into a column, will be much less cumbersome, and much more flexible for tailoring directivity to just what you may need. If you have a column containing seven such units in the vertical direction (assuming you stand the column up),

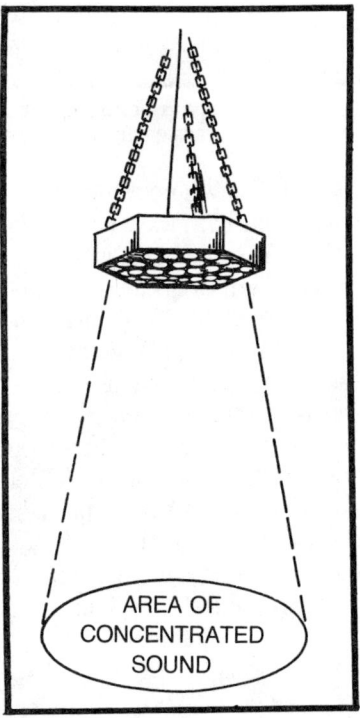

Fig. 7-3. How a beaming loudspeaker can be used to serve a limited area with sound.

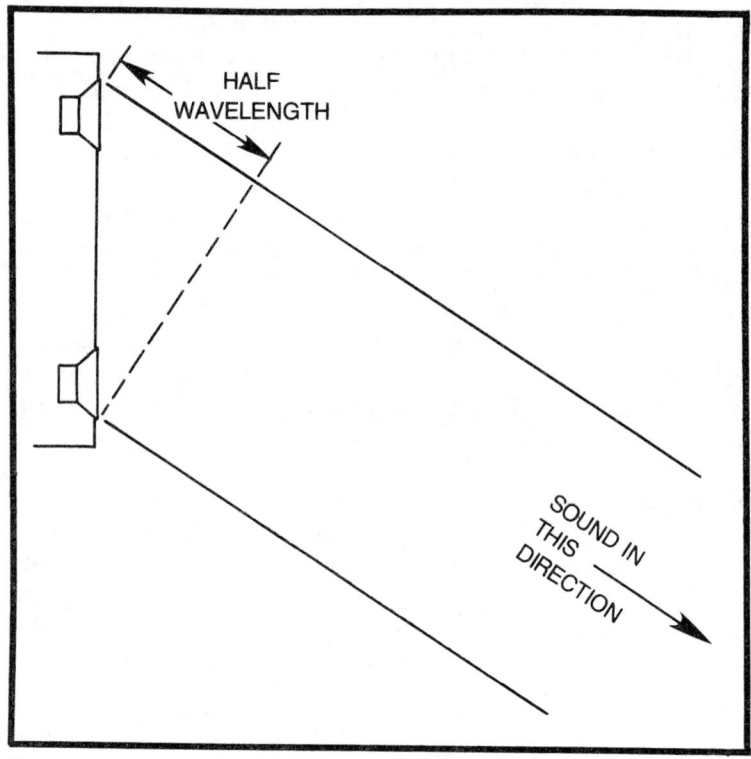

Fig. 7-4. At an angle to the column that makes the difference in distance from its two ends a half-wavelength or more, cancellation prevents radiation of sound.

the radiation will be as directional as a bank of units that measures seven units each way. But in the horizontal direction, or plane, the sound will spread out, just as it would from a single unit.

Just as the outside ring of units move more air than those nearer the middle of the 37-unit assembly, so in a column the end units will move more air than units toward the middle of the column. But in this case, the middle units also move air at the sides, which is why the sound radiation fans out sideways.

It does not fan out at the ends, because sound from the other units produces cancellation in that direction. If you consider a direction (Fig. 7-4) such that from a distance a unit at one end of a column is half a wavelength further away than a unit at the other end, one of these units will be pushing when the other unit is pulling. This means there will be little resultant sound in such a direction.

Thus a column loudspeaker produces a cylindrical wave, with the column as its axis. Very little sound radiates beyond the ends of the cylinder (Fig. 7-5). This does not duplicate any natural source of

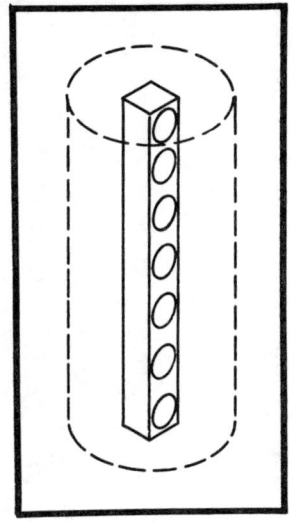

Fig. 7-5. One way to visualize how a column loudspeaker works is to realize that it radiates a cylindrical wave.

sound, so when you are in a room fed with column loudspeakers, you will probably find it difficult to pinpoint the source of sound. But because it avoids unnecessary reflections when properly used, it will be good sound that is easy to listen to.

This means that placing them vertically, perhaps with a slight tilt forward to aim at a seating area built on a slope, will concentrate the sound to within the seating area with none of it wasted. At the same time the sound will fan out horizontally to cover all the seating area (Fig. 7-6).

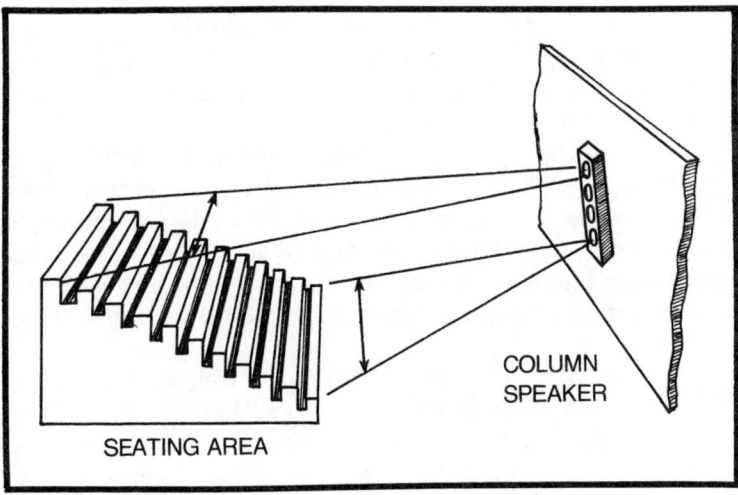

Fig. 7-6. Using a column loudspeaker for a similar purpose.

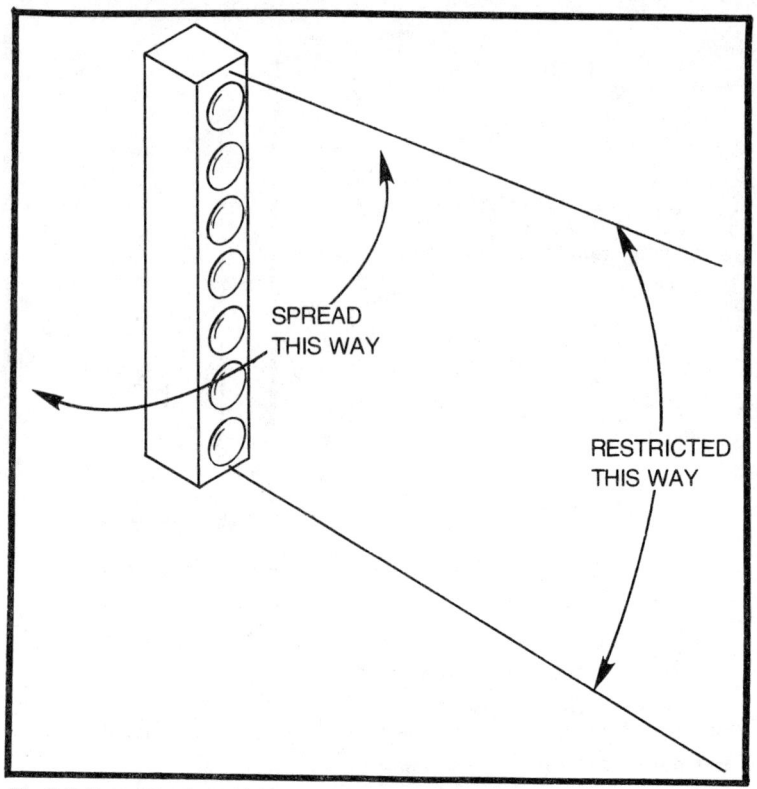

Fig. 7-7. Remembering which way a column speaker concentrates its beam, and which way it spreads the sound.

This method can result in considerable reduction in confusion in rooms that have proved very difficult previously. It also explains the tremendous popularity that professionally made column speaker units have enjoyed in recent years.

Once you appreciate how this principle works, you will be able to see situations where it will help get good sound performance not before possible. But you have the same psychological mistake to guard against, that we mentioned relative to oval units. The direction in which the sound fans out is crosswise to the direction of the column, and restricted relative to its lengthwise dimension (Fig. 7-7).

Suppose you have a long, narrow room with highly reflective walls and a fairly low ceiling. For this you might find it better to lay the column down and place it against the ceiling (Fig. 7-8). Being against the ceiling will avoid ceiling reflections. If the seating is occupied, sound reaching the floor will be absorbed by the audience.

The sound that would cause confusion in such a room would be that bounced off the walls, which laying the column down prevents from happening. Being a long room, reflection from the back wall is probably less important. Incidentally, the place where reflection from the back wall will produce the worst effect is about two-thirds of the way back.

Right at the back wall, the reflection has too little delay behind the original sound to affect the listening intelligibility. At the front, the listeners are near enough to the loudspeaker that the reflection all the way from the back will not bother them. But about one-third of the way back toward the front, after reflection has occurred, the interference due to that reflection will seem at its worst.

The best thing to do, is probably to find some means to stop the back wall reflection, although there are more sophisticated solutions that we will not discuss in this book.

In some sound studios, absorptive shapes are used. These are wedge-shaped pieces made of material that is moderately absorbent. However, because it is also partially reflective, using such odd shapes means the reflections do not make up a consistent echo, thus the original sound becomes stronger compared with the reflected sound.

Curtains can produce some absorption also. But they must be fairly heavy and, even then, they affect the higher frequencies far more than the lower ones. To absorb low frequencies, curtains would have to be extremely thick and correspondingly heavy. These are solutions you may want to try in some instances.

The problem with such solutions usually is that in any room big enough to cause that kind of a problem, the area to be treated is

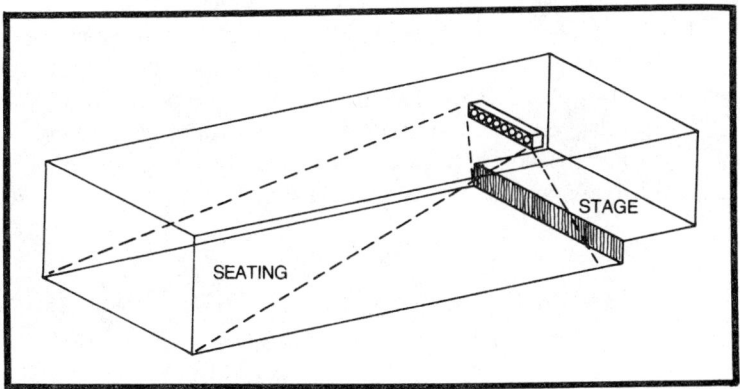

Fig. 7-8. Another way of placing a line radiator (column) to get a desired coverage, without unwanted reflections.

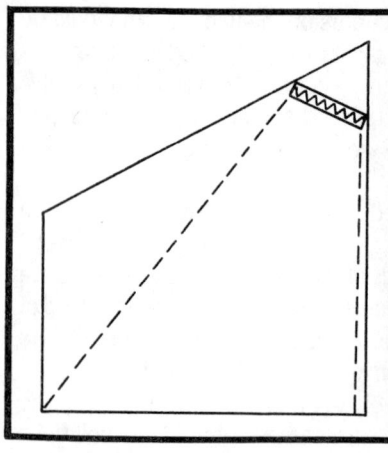

Fig. 7-9. Using the line radiator, or column, to get good performance in a room with a cathedral ceiling.

large, and the cost of doing it is thus too high to be feasible. If similar results can be achieved by careful loudspeaker selection and placement, it will be a lot cheaper.

We already mentioned the grandstand at the racetrack, where this kind of system did a good job. If you have a room with a cathedral ceiling, you may find it advantageous to do the same thing. Here you want to avoid reflections off the sloping ceiling and also the taller wall. So how would you place the column speaker?

Crosswise, near the apex of the ceiling (Fig. 7-9) is the best answer. One at or near each end of this apex would make for good stereo reproduction in the area below. If you put more conventional units at floor or table level facing the rest of the room where people normally sit, sound can go up into the cathedral ceiling part, and bounce around, creating an effect which might be all right for some programs, but could be quite distracting on others.

For quadraphonic reproduction in such a room, the remaining two channels could probably use omnidirectional speakers, because their purpose is usually to add controlled reverberation anyway.

In all of this, the important thing is to understand what you are doing, and why, based on the factors we have studied in the earlier part of this book, so that you can do it right.

8

Putting Together More Sophisticated Types

To understand how to put together more sophisticated loudspeakers, we need to see what such units are capable of doing. Let us start with the prototype systems from which most of them were derived with certain variations: the infinite baffle and the bass reflex.

For those prototype units, ordinary loudspeaker units were used because the idea of designing a unit specially for an enclosure to go with it had not been thought of in those days. Back in those days, each part of the job was tackled separately, one step at a time.

PROGRESS IN LOUDSPEAKER DESIGN

In those days, the first step was to design the best possible loudspeaker units in a whole variety of suitable sizes. Then various boxes, or enclosures, were designed in which to put them to get the best performance out of them. That was the only procedure followed for a long time. Let us briefly review how it went with each kind.

You start with a loudspeaker unit, whose diaphragm resonance, unmounted—that is, with it held out in space with nothing close to it—was about 110 hertz. That was a typical value, some were lower, some higher.

This resulted from careful design of the unit. It had the lightest diaphragm that could be made with a consistent degree of rigidity, which the shape helped to achieve. The shape of the diaphragm was conical and perhaps had ridges pressed or formed in it, coupled with a suspension that controlled its movement effectively, while allowing it the maximum freedom of back and forth movement.

An important function of the concentric ridges pressed into most loudspeaker cones—one that is particularly important with the larger cone sizes—is what we may call *successive decoupling*. This behavior occurs due to two facts: (1) that increasing frequency causes all of the cone, with the air in contact with it, to offer more resistance to movement due to its mass, and (2) that increasing frequency also requires less movement (that acoustic engineers call *displacement*) to produce a corresponding sound pressure.

So what happens is this: at resonance, the whole cone, with the air in contact with it, moves as a single mass, tuned with the compliance of the suspension, which will include the air cushion inside the box if the box is closed. That is close to the lowest frequency the unit can successfully reproduce. As we have pointed out, this is not a fixed frequency, because it can be affected by the kind and size of box used to house the unit.

Above resonant frequency, because higher frequencies want to move the cone and the air it drives at a faster rate, the mass of the cone with the air it drives produces increasing opposition to that movement, which is greatest where the area is greatest—around the outside of the cone. This means that if the outer rings can be progressively released from having to move with the inner ones driven by the voice coil, the inner ones will be able to move more freely.

So as frequency increases, by providing the cone with these concentric rings, the whole cone moves at resonant frequency. Then the outer ring ceases to move as much as those inside it. As frequency rises further less of the cone moves until only the innermost part, adjacent to the voice coil, is moving.

The rings, or concentric ridges, are only slight. This means that there is no sudden point in frequency at which part of the cone ceases moving because there is no definite flexing or nonflexing demarcation. The secret of good loudspeaker unit design is to select material and thickness of which to make the cone, as well as the shape of the cone, so as to get the smoothest possible overall frequency response.

Such design work resulted in a unit with a resonance around 110 hertz, enabling them to give as good a performance as possible over as much as possible of the audio range. Larger units would have a somewhat lower resonance, smaller units a somewhat higher one. Big units of those days might resonate as low as 90 hertz, small ones might go as high as 150 hertz. Around 110 hertz was generally regarded as a good figure.

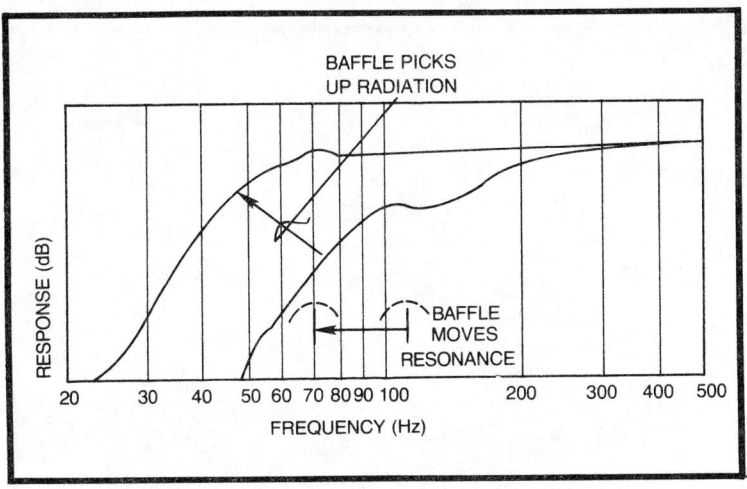

Fig. 8-1. How putting a unit in a baffle improved its low-frequency performance.

Now, you take that unit and put it in a large baffle board. That will make its resonance drop from 110 hertz to somewhere between 60 and 80 hertz. This happens becuase putting it in the baffle forces the diaphragm to move more air at the lower frequencies. You are effectively adding weight or mass to the diaphragm, thus tuning it to a lower frequency.

But the improvement in bass response, resulting from putting the unit in a baffle, is more than just lowering the resonance would suggest. This is because you are getting the low-frequency waves out into the room better, instead of allowing them to shuffle round the edges of the diaphragm (Fig. 8-1). This improvement is what made baffles so popular, when they were first introduced. Also, of course, they were an easy do-it-yourself project.

But they were awkward, and the back of the unit was unprotected, so designers looked for ways to put them in more convenient boxes. If you fold the edges of the baffle back, but leave the back open, the open-back box resulting behaves very much like a baffle of somewhat larger dimensions than the box. Putting the loudspeaker unit in the hole in the front of the box lowers the resonant frequency in the same way, and improves the bass radiation too.

The open-back box, like the baffle, had one disadvantage. Perhaps it was more noticeable with the open-back box. If you sit directly in front (Fig. 8-2A) the response is at its best. If you sit at the back, the bass may be as good, but the high frequencies are missing (curve C). But if you sit at the side, the low frequencies are badly missing, and the high frequency response is erratic (curve B).

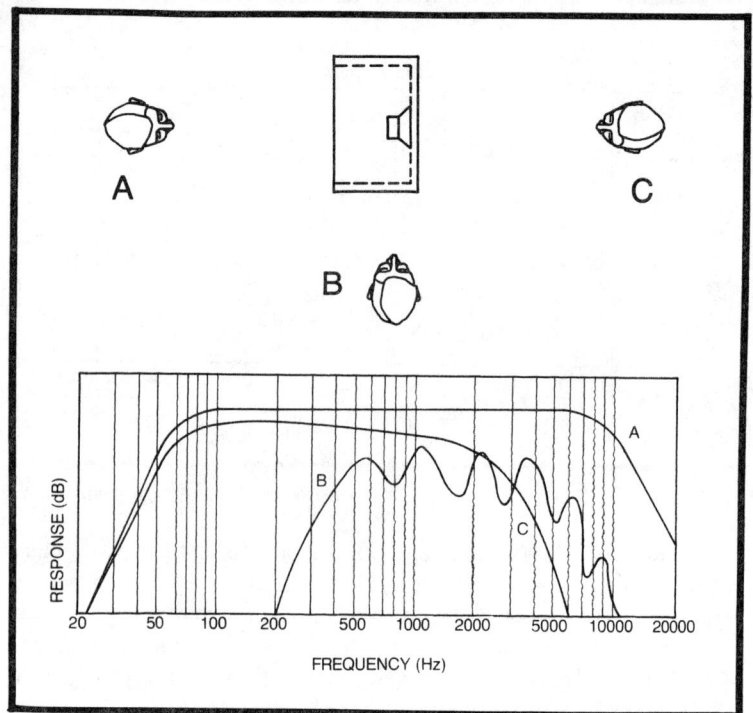

Fig. 8-2. Different responses associated with an open-backed loudspeaker, according to where you listen to it.

At the back, the low-frequency response is as good as it is sitting at the front (curve C), but the high frequencies do not get out because that section of the cone has no direct access to the air behind. The waves have to find their way past the obstacle presented by the back of the loudspeaker unit.

We have discussed just three typical listening positions, relative to such a loudspeaker. But in everyday use, people do not sit in such closely defined positions. If you asked most people to sit in such a position, they would probably ask, "Do I have to?" with the implication that by designing something that requires them to listen from such mandatory positions, the designer is infringing on their civil rights.

This was in fact a problem with the early stereo demonstrations: the way the program material was prepared, that is, the differences between what was put on the left and right channels, made it necessary for the listener to sit at a position equal in distance from the two loudspeakers (Fig. 8-3) if he was to get any appreciable stereo effect at all.

So further work on stereo systems sought to remove this limitation, both by working on the way left and right were separated on the channels, and by working on better loudspeaker systems to convey that information to the ears of listeners, which is what this book is about.

But, with reference to listening to an open-back loudspeaker, the almost continuous change in quality of sound, as you move to different listening positions, made it undesirable, even for monophonic listening. For stereo, it becomes impossibly difficult. The only place where it goves good reproduction is in front on axis.

This means that for good stereo such a loudspeaker type being used would require, not only that the listener sit at equal distance from each unit, but that the units be turned in to face the spot where the listener is sitting. So you can see, good stereo listening with this type would require the listener to sit in a sort of *third-degree* position with everything focused on him.

Earlier in the book, we spoke of using flat baffles, particularly in pairs, for stereo. But notice that in such systems we avoid feeding the low frequencies, the ones that drop off at the side, into these

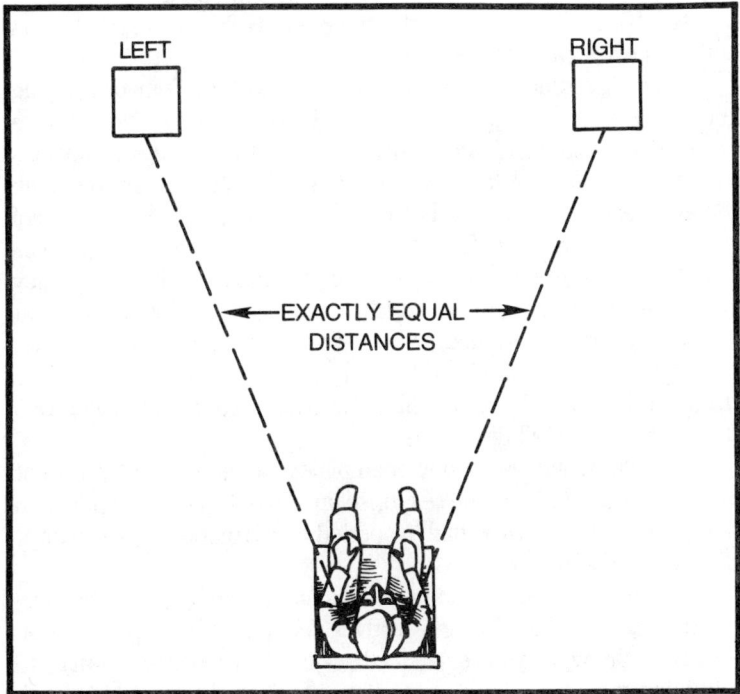

Fig. 8-3. The special position required for listening in the early days of stereo.

115

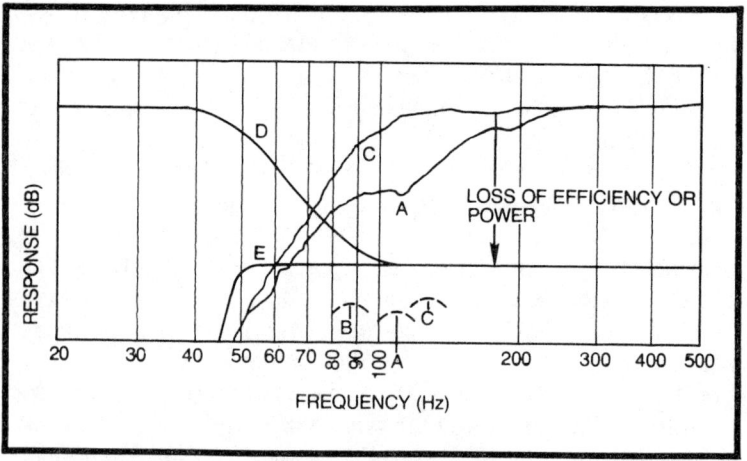

Fig. 8-4. Compensating an infinite baffle loudspeaker to get more bass response. (A) Free air resonance of unit. (B) Resonance in box without back on. (C) Response and resonance with back on. (D) Electrical response compensation. (E) Overall response with compensation.

units. We feed these low frequencies into a single woofer unit, whose job is to fill the room with components of the sound that are in this frequency range.

That is a much more sophisticated system than developing technology was ready for at this stage. When open-back boxes were the next step from flat baffles, the solutin to the low-frequency loss at the sides was to enclose the back. It is true that this improved the low-frequency response at the sides. But it did so at the expense of the response in front, in fact of the overall low-frequency response.

What closing the back in did, was to make the low-frequency response uniformly good in all directions instead of having that tremendous difference that the open-back variety had. By this time, electronics was coming into its own, so the loss of bass could be made up electronically. Then this infinite baffle, as it was called, was uniformly good in all directions.

But this meant bigger power amplifiers to feed enough power at low frequencies to keep the response up. To put this development in perspective, notice what had happened to our good loudspeaker at this point (Fig. 8-4).

The unmounted unit had a resonance at about 110 hertz. Putting it in the open-back box lowered the resonance to maybe 80 or 90 hertz (B). But when you sealed the back in, the resonance went up to a frequency higher than it started at, possibly 130 hertz or higher (C). The smaller the box, the higher the resonance goes, and the frequency response drops off badly below whatever frequency the resonance goes to.

Then, to get the bass response back, electronic equalization is applied (curve D) so the response goes down to some lower cutoff (curve E). But doing this requires a lot of extra power, needed mainly to drive the unit down to this lower frequency.

Make sure you see what the infinite baffle did, and how a good response was obtained with it. This will help you see how the acoustic suspension differs from it. It used an ordinary loudspeaker unit that might be used for any type of enclosure. It was just a good, general-purpose unit. So at frequencies high enough for the acoustic cushion inside the box to have any effect, the infinite baffle was as efficient as any other enclosure.

But it had this bad drop off at low frequencies. Many infinite baffles were used that way, which means they had very limited low-frequency response. For an office intercom system, or a paging system, or background *music while you work*, that is quite adequate: you don't want that booming bass—it can be distracting.

On the other hand, infinite baffles can be pushed by electronic equalization, as shown in Fig. 8-5. But this involves extra power, a lot of it, especially at the low frequencies. So we have a system in which the power demand at midrange and higher frequencies is about normal for that kind of unit, but in which power demand at low frequencies may increase between 10 and 100 times.

The only thing that made this approach feasible at all when it was used was the greater ease with which power could be obtained as amplifier technology improved. In tube amplifier design, the advent of beam tetrodes in large sizes, then when solid state came

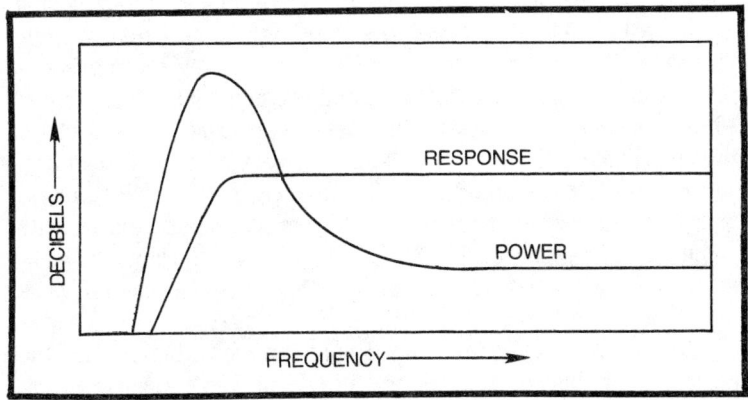

Fig. 8-5. How electrical equalization of an infinite baffle loudspeaker requires extra power at the low frequencies.

along, even more efficient output devices of the power transistor type.

But solid state came along a little too late for that. The acoustic suspension and loaded reflex designs were already in existence before economic solid-state high-power amplifiers became available. However, for the moment to get our perspective, let us revert to the time when the latest loudspeaker type was the infinite baffle. Do you get the picture, as of that time?

THE PROBLEM

So our acoustic designers got to thinking. Of course, the advocates of horn loudspeakers were still making them, and claiming they were the best, in fact the only *real* loudspeaker. But not too many people listened to that claim, because they were just too big, for a great many people's living rooms.

What the designers wanted was something that would use a more reasonable sized box, and yet could reproduce the bass frequencies more efficiently than the so-called infinite baffle.

The problem with the infinite baffle was that sound energy from the back of the diaphragm was not only being wasted, the box was, in fact, absorbing more than its share of energy just to keep the energy from the back from getting out. The idea that started the whole bass reflex revolution was simply to shift the phase of the back radiation acoustically so it would add to what came from the front, instead of canceling it.

Putting a port, vent, or whatever you want to call it, alongside the loudspeaker unit produced an acoustic filter system. The air in back of the box acts like a spring, while the air in the vent acts like a weight. The two together can delay the movement so the air moving in the vent goes forward when the diaphragm moves forward, and vice versa (Fig. 8-6) at the lowest frequency the system radiates.

Thus the whole family of bass reflex designs, which constituted a big step forward, was born. It used a new principle. So far we have the open baffle and the open-backed box, very little used anymore. The ideas put forward in Chapter 6 for utilizing built-in acoustic organ pipe features had been toyed with by earlier experimenters with some interesting successes. And, of course, the horn used *proper* sound wave propagation even though the units that used it were much too big.

But the bass reflex introduced a new principle, built around the acoustic filter concept. A horn, or a tuned pipe, uses sound propagation concepts based on the fact that sound waves take time to travel. That fact is what involves designers in big dimensions. Sound travels

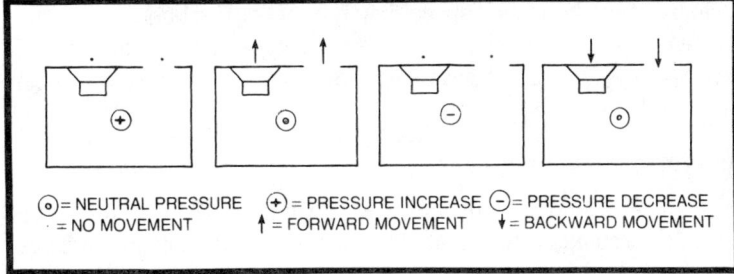

Fig. 8-6. A sequence diagram, showing what happens at resonance in a bass reflex box.

in air at more than 1000 feet per second, which makes the wavelength, particularly of low frequencies, quite large. And designers of systems that use these facts, are always involved with half-wavelengths, quarter-wavelengths, or some other dimension which inevitably turns out to be large.

The bass reflex principle got away from that, by separating the properties of air on which propagation depends: mass and elasticity. It uses what acoustic engineers call *lumped parameters*. They are not basically interested in waves propagating across the inside of the box, although that can cause them trouble, as we have mentioned elsewhere, if dimensions are not carefully chosen.

In the bass reflex enclosure, the air inside the box becomes just a compliance (as it is incidentally in the infinite baffle also). The air in the port or vent, or the dummy cone similarly installed, is a mass. The two interact as separate, virtually lumped parameters to produce the reinforcement for which the bass reflex is noted.

This basically is what makes the bass reflex break away from the limitations that had seemed impenetrable up to that point in time. Until then designers had labored and labored, hoping to find a way to get more bass for smaller boxes. And until then there seemed no real solution. And of course the same limitations still apply. It is just that later developments have found other ways to get around them.

This trick enabled more bass to be obtained from the same sized box. The resonance was no longer pushed up so far in frequency be sealing the box, so everything was better. But still the size of box needed to get really good low-frequency response seemed unreasonably big to people with small living rooms, such as apartments.

This problem became more acute as stereo gained in popularity, requiring two speakers where the old hi-fi (mono) managed with only one. As every apartment seemed to have a bookshelf, often including more bookshelf space than the owner had books to occupy it, the logical idea of designing an adequate loudspeaker that would fit on a bookshelf pressed many designers to work at finding a way to do just that.

ACOUSTIC SUSPENSION

The first kind of loudspeaker to break this size barrier was the so-called acoustic suspension. This starts with the infinite baffle as its prototype, but overcomes the low-frequency problem by redesigning the unit so it will work only in the box designed to go with it.

The unit, unmounted, has an extremely low resonant frequency (Fig. 4-14) that could never be used. In fact to try to do so would destroy the unit as a rule. But when the same unit is placed in its properly sealed enclosure, the resonance comes up to a frequency around 30 hertz, controlled by the heavy diaphragm and the acoustic stiffness of the air inside the box.

This was the first time the loudspeaker design had been tackled as a whole. Previously, one group of engineers designed the best loudspeaker units they could with slightly varying parameters, such as resonant frequency, size cone, and so forth. And another group of engineers designed enclosures of one kind or another in which to put these more or less standard units.

With acoustic suspension, the loudspeaker unit was modified very considerably, being designed only to be used with a certain sized box. Before that, interaction between the two had been something designers felt they had to get along with, one way or another. This was an attempt to use that interaction to do something quite different. And it worked.

Prior to the advent of acoustic suspension, any attempt to get enough output at low frequencies from such a small box required driving the unit so hard, with so much diaphragm movement, that it produced considerable distortion due to nonlinear restoring force on the diaphragm. Such excessive movement meant that the force pushing the cone back to its normal position was no longer proportional to its deflection from that position. Because of this nonlinearity, such overdriven loudspeakers produced a lot of distortion.

To understand this, you need to see why loudspeakers distort at low frequencies when the movement gets large. There are several reasons for this. The first could be a nonlinear driving force. An ordinary loudspeaker unit has a voice coil that just fills the magnetic air gap in which it moves back and forth (Fig. 8-7). When it moves any appreciable distance, it leaves the gap. When the turns of the coil are no longer in the gap, the drive force is reduced.

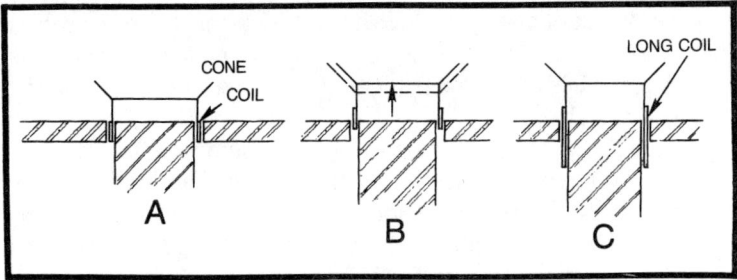

Fig. 8-7. How acoustic suspension loudspeaker units improve linearity of drive. (A) Section through a normal unit magnet and voice coil. (B) The normal coil driven one way as far as would be necessary to acoustic suspension, almost leaves the magnetic air gap, losing its drive so the force becomes nonlinear. (C) The acoustic suspension unit has a long coil so that when part of it moves out of the gap either way another part of it moves in, making the drive much more linear.

To overcome this, a longer coil is used in the specially designed units for acoustic suspension use (also shown at Fig. 8-7). The result is that as part of the coil moves out of the gap, another part moves in, and there is always the same amount of the coil in the active part of the gap. This linearizes drive force for large movement such as the acoustic suspension type employs.

The second cause of nonlinear output, which is distortion, is the suspension. Linear suspension would provide a restoring force, tending to bring the coil and cone back to their position of rest, that is precisely proportional to the distance by which it is moved from that position of rest. For small movements, this is true of almost any suspension. But for large movements, the cone reaches a point where the suspension sort of gets tight. Any further movement will require much more than a proportionate increase in force to produce it.

The way acoustic suspension tackles this is two-fold. If you removed the mechanical suspension, just letting the coil and cone move back and forward as far as it wanted without any control, when it is put in the sealed box the compression and expansion of the air inside the box would be the only restoring force the cone would have. And compressing and expanding a volume of air by a little bit is a lot more linear than any mechanical restoring force could be.

But it is not practical to make the air the only restoring force. If you did, then expansion due to temperature rise (because the room got warmer, or barometric change in pressure with the weather, would cause the cone to pull or push the coil completely out of the gap. So a small mechanical restoring force that will bring the cone to its middle position when it is not being driven either way is essential.

The result of making the mechanical restoring force so small is that, when the air cushion in the box is not there, the movement is very floppy. Only when you put it all together does it work how the designer intended.

Acoustic suspension almost completely removes the mechanical restoring force, and substitutes the cushion of air inside the box for this purpose. The restoring force provided by the cushion of air is very linear by its very nature. So this kind of loudspeaker produces much less distortion than earlier loudspeakers of similar size when working at a similar acoustic output.

The acoustic suspension was the first step toward low-distortion, small-box loudspeakers with good bass response. But the cone still traveled as far as it did in an infinite baffle of the same size; however, it was better controlled to get the lower distortion. So, although acoustic suspension solved some problems, some remained.

Because the cone was so heavy, the electrical output applied to it spends far more energy moving the cone than it does moving the air that makes the sound wave. So it is very inefficient. Where the compensated infinite baffle lost efficiency only at low frequencies, the acoustic suspension becomes similarly inefficient at all frequencies.

And because it will not put out much sound at higher frequencies, it must be used with midrange and tweeter units to supply those frequencies. This inefficiency means that to get an output from an acoustic suspension loudspeaker, comparable with earlier loudspeakers capable of similar quality (but larger), requires between 10 and 100 times the electrical power to drive it.

While acoustic suspension does get lower bass into much smaller boxes, we have not entirely defeated the principle that you need bigger boxes to get more bass. Acoustic suspension achieves it at the cost of efficiency. The smaller the box, for a given bass response, the less efficient must the loudspeaker be.

Making the box smaller makes the acoustic suspension that much stiffer. And that means we must make the mass of the moving system, the coil and cone, correspondingly heavier to keep the resonant frequency from moving up. So, if you keep the bass response the same, making the box smaller requires the cone to be heavier, or the coil, or some part of what moves. And this means that more energy will be used moving the cone than moving the air that makes the eventual sound wave.

So smaller boxes with the same bass response must have lower efficiency by their very nature. We have sort of cheated on the infinite baffle principle to make the box smaller, as much so as we want, but it can only be achieved by making the moving system heavier, which costs a corresponding loss in efficiency.

LOADED REFLEX

Just as the bass reflex helped as a step forward from the infinite baffle, applying the acoustic suspension to the bass reflex resulted in another compromise. Various proprietary names have been given to this, but we will give the generic name of *loaded reflex*.

It starts with a loudspeaker unit, in which the resonance, unmounted, has been shifted down in the same way as in units for the acoustic suspension type, but usually not so far. The fact that the major stiffness encountered by the cone movement as it approaches resonance, is due to the air in the box, not to its mechanical suspension, means that is where a substantial part of the energy goes.

This energy is still inside the box. In the acoustic suspension unit, it never gets out of the box. The loaded reflex differs in that it uses a vent or port, or a loaded diaphragm in the port, to convert this energy into radiation. In fact, where this design differs from the old bass reflex, is that much more sound comes from the secondary source of output at the lowest frequency.

At resonance, most of the energy supplied to the voice coil that drives the main cone is spent compressing and expanding the air inside the box with not too much movement—not as much as would be needed if it were an acoustic suspension type. Now the pressure fluctuations inside the box produce a much bigger movement of the loaded diaphragm, or the air in the port, than the main cone shows.

This means that less energy has to be spent driving the somewhat heavier cone of the driven speaker unit. Thus this approach results in a more efficient design than the acoustic suspension for an equivalent size, just as the bass reflex did over the infinite baffle in earlier days.

DESIGNING YOUR OWN

Now comes the question of how you can use these facts in designing your own system. In the first place, manufacturers who make such units, supply designs for cabinets to go with them because the units themselves will not work like any ordinary speaker will. You can always follow these designs and be assured of good results.

What we want to do here is to help you understand why the results work so that, if you want to change something about the design, you will know what you are doing and will not make some silly mistake, thinking it will make no difference when in fact you are altering something critical.

With the acoustic suspension type, the box has only one variable: its contained volume. If you make the box smaller in volume, that increases the stiffness the air imposes on the cone, raising the resonant frequency, thus the lowest frequency that the loudspeaker will reproduce.

This will follow a strict relationship. If you halve the volume, the frequency will go up by a factor of about 1.4 times, and so forth. It is not advisable to make the box bigger than the design value provided because that could allow excessive movement which might damage the unit.

When we speak of halving the volume, or doubling it, do not think that means changing all the dimensions by that factor. Just changing one dimension will do it, after allowing for the space occupied by the unit itself. If you make each dimension 4/5 of the original dimensions, the volumme will be reduced by the cube of this, or 64/125ths, which is close to half. And that will put the cutoff frequency up by the square root of 2, or 1.4. If the original unit had a low frequency cutoff of 35 hertz, the reduction of all dimensions to 4/5ths will raise the low-frequency cutoff to about 50 hertz.

With the loaded reflex type, as well as the enclosed volume of air, we have the characteristics of the vent or duct, which give it at least two more variables to think about. In scientific terms, the tuning of the vent is determined by its mass, which means the equivalent weight of air moving against the stiffness of the volume of air contained inside the box.

The further complication occurs because, in acoustics, such mass is not a simple dimension that might be measured in pounds or ounces, or some equivalent measure for weights. The area comes into the picture as well, in at least two places (Fig. 8-8). First, on the inside of the vent, where it meets with the volume of air inside the box, the area of the vent determines how much of the energy, represented by the compression and expansion of air inside the box, gets coupled to the air movement in the vent. The bigger the vent, the better the coupling.

Let us see what this means. Suppose the box is a foot cube for convenience, although that is not a good shape to use, but it is easy to visualize. Now suppose the vent is a tube, a quarter-inch in diameter. Fairly obviously, most of that cubic foot will not know the vent is there. Only a little air, near the vent, will go in and out.

On the other hand, if you have a hole in one side 8 inches square, that will take up most of that side, 64 square inches out of

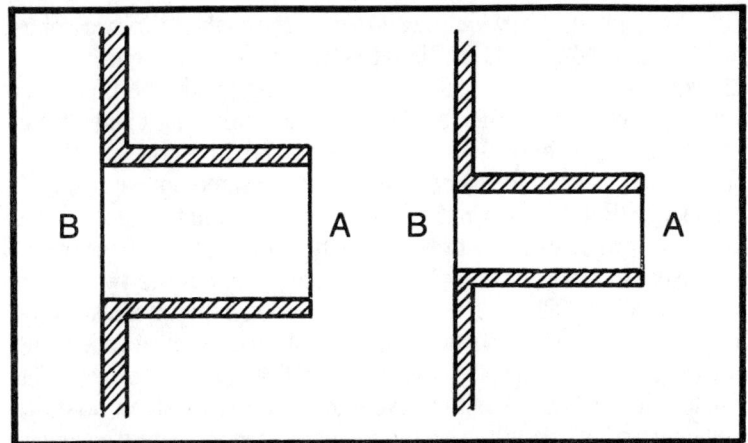

Fig. 8-8. The important places in a duct, used in a loaded reflex: A. where the duct couplers to the air inside the box; B. where the duct radiates its component of sound.

144 square inches. Air will rush in and out too easily, and the resonance will be indefinite. Somewhere between those extremes is a size that will give the best resonance for your purpose.

The important thing is that the sizes must be such that the air inside does little else than change pressure, while the air in the vent does little else than move back and forth. But don't take that to extremes, like the quarter-inch vent to a cubic foot box. True that would make a good resonator with quite a sharp tune point, like a Helmholtz bottle. But it would not be an efficient energy converter.

The extreme condition, on which the Helmholtz resonator is based, produces a very sharply tuned resonance. You might indeed get a good sound output, when the reproduced note hit the resonant frequency exactly. But if the frequency reproduced is half a semitone away from resonance, the output would disappear.

The opposite condition, like the 8-inch square hole in a foot cube box, would produce a coloration not noticeable if the side where the 8-inch square hole was, is made completely open. But it would not pull up the response where you need it.

When you have the right proportions, the resonance produces enough increase in response over a broader band and delivers the energy as sound that the cone itself cannot because it is too heavy. In short the loaded reflex behaves like a sort of tuned transformer to step up the energy by properly matching it to the air into which it radiates over a range of frequencies including resonance. To work right, the resonance we are talking about includes both the cone and the vent, or loaded duct.

Remember how we showed this worked when the two cones were identical (Fig. 4-11)? There the two cones move equally and in the same direction at resonance, with the air inside the box being compressed and expanded as the two cones move back and forth in unison. In the loaded reflex, both cones, or the drive cone and the air in the duct, move in the same direction at the same time at resonant frequency. But the cone in the opening, or the air in the duct, moves more, or produces most of the acoustic output at this frequency.

Near to resonance a similar thing happens, but the two movements gradually fall out of step. An interesting experiment, to show how it works, can be devised using an oscilloscope with a cone in the loaded port that has a voice coil. But the voice coil is not driven. The voltage it develops due to movement is used to show how the relationship between the two movements changes with frequency. Most readers will not have this opportunity, so here is how it looks.

As we said, at resonance, the two move together, in the same phase, synchronized as it were. At a frequency below resonance, the loading cone moves a little earlier than the drive one. At a frequency above resonance, the loading cone moves a little later than the drive one. And on both sides of resonance, the movement of the loading cone reduces, while that of the drive cone may increase a little.

Whether the drive cone increases its movement away from resonance depends to some extent on the amplifier internal impedance, which we discuss later under *Damping Factor*. If the damping factor is high, the movement of the drive cone will not increase away from resonance, but the loading cone movement will reduce a little more.

And the size of the vent where it emerges into the room, the outside world, determines further how much energy it can radiate, at the frequency where resonance occurs, out into the room.

From the viewpoint, only of determining frequency, you have two factors: the stiffness of the air in the box, and the mass of the air in the vent. The stiffness of the air in the vent depends on the ratio of length of the duct or vent to its area.

To think about this mass, first there is a minimum length physically, where the opening behaves as a hole rather than as a tube. Here you have to consider the air approaching and leaving the hole, as well as the small part that is actually in the hole (Fig. 8-9).

If you use a round hole, fairly reliable figures can be given for this minimum effective length that depend only on the radius of the hole. If, as is more usually the case, the hole is rectangular, the calculation is a little more involved. Then you take the area of the

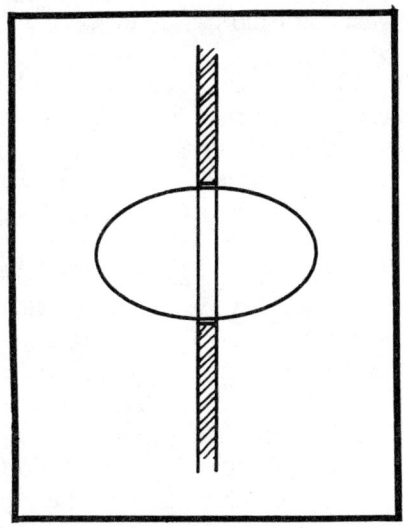

Fig. 8-9. Concept of air associated with an ordinary port, or unducted hole.

hole, divide by π(3.14), and extract the square root to find an equivalent value for the radius.

Then, with a hole that has no duct extending from it, the equivalent length, due to air coming and going, is proportional to its size measured as a diameter or radius, while the amount of air that can flow depends on the area of the hole. Considering just the effective mass, for the purpose of calculating the tune point, it will be inversely proportional to radius or diameter if it has some other shape than round and is changed in size without changing the shape.

What this means is that doubling the dimensions will halve the mass, or halving the dimensions will double the mass, as it affects the tuning of the energy coming out of the hole.

Now, if a duct is connected to the hole, then the length of the duct becomes important. The mass is proportional to the length of the duct, after allowing for the equivalent minimum length when it is physically only a hole without any measurable length. Let us use some figures to illustrate.

Suppose we start with a hole that is 4 by 8 inches. That makes an area of 32 square inches. Divide by π(3.14), and you have approximately 10 square inches. Now take the square root, giving as your equivalent radius about 3 1/16 inches. Now, the equivalent minimum length of such a hole can be taken as about 1.7 times its equivalent radius, which figures to about 5⅜ inches.

What this means, is that to double the equivalent mass of this hole, would require a duct that is an additional 5⅜ inches long. (Fig. 8-7). Similarly, doubling the actual duct length to 10¾ inches long

127

would make the mass three times that of the same sized hole without any duct.

Now, suppose we make the hole smaller, half the size each way, 2 by 4 inches. Following the same method of calculation, the equivalent radius will also be halved, to about 1 7/12 inches and the equivalent minimum length will also be halved, to about 2 11/16 inches. Its equivalent mass, just as a hole, will be twice that of the larger hole.

Now, adding a duct 2 11/16 inches long will make the equivalent mass twice that of the smaller hole, or four times that of the larger hole. Making the mass four times will just halve resonant frequency. So changing from a 4- by 8-inch hole with no duct, to a 2- by 4-inch hole with a 2 11/16-inch long duct, will halve resonant frequency.

That sounds like a good trick until you realize that halving the dimensions makes the area one-forth, so that only one-fourth as much sound wave will come out of the smaller hole, other things being equal. Do you begin to see the problem? What would be better, would be to increase the mass by lengthing the duct, but at the larger size, the duct would need to be three times the equivalent minimum length of 5⅜ inches, which is 16⅛ inches long.

Make sure you understand this concept of *equivalent minimum length*, because it makes the calculations much easier, if not quite foolproof. If you overlook the air that moves on either side of the hole or duct, when it is in fact only a hole, your calculations will be off. The 4- by 8-inch hole is equivalent to a round one of about 3 1/6-inch radius. And our formula shows that a hole of this size has an equivalent minimum length of 1.7 times this, or about 5⅜ inches.

This means that, when the 4- by 8-inch hole has only the thickness of the wood or other material of which the box is made, it has an equivalent length of 5⅜ inches. So, when you lengthen it by attaching a 4- by 8-inch duct, that is added to this minimum equivalent. If you add an actual 5⅜ inches of duct, then the equivalent length of duct is 5⅜ inches equivalent minimum plus 5⅜ inches actual, making a total of 10¾ inches.

And to get an equivalent duct length of 21½ inches, you subtract the equivalent minimum of 5⅜ inches to get an actual duct length of 16⅛ inches, which is three times the equivalent minimum length.

Any duct finishes up being a compromise in size and length. It must tune the box to the right resonance frequency. But if you make it too big, it takes too much away from the volume of the box, and defeats its objective that way. And if you make it too small, you lose by having its turning too sharp at resonance, and by not converting

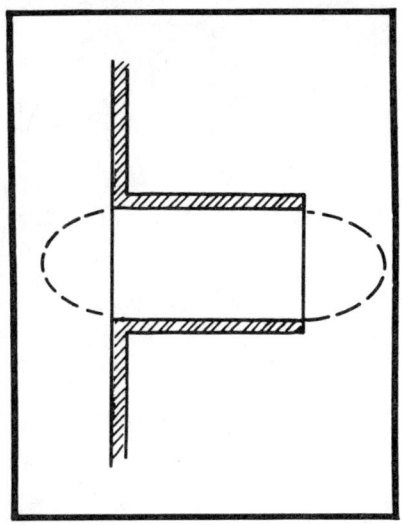

Fig. 8-10. How a duct increases the amount of "movable air" associated with the reflex action.

enough energy through the duct to produce the boost in output over this range of frequencies that you are looking for.

You may find it difficult to accommodate that much duct inside the box. You can perform some more calculations to find a better way of tuning your system if you want to, but it will usually be easier to merely translate what the manufacturer supplies as data to a little different form that suits the shape you can use.

This means you should keep the same duct or hole area, and also the same length, as well as the same volume inside the box. And in doing this, don't forget that you cannot count the volume occupied by the duct as part of the volume inside the box. Also that the inside end of the duct must provide free access—very free access—to the air volume inside the box.

9

Things That Can Ruin Your Performance

Assume you have used the information given in earlier chapters to make some changes in your system, or to adapt it to your needs, and now it does not work quite how you expected. What did you do wrong? Maybe more important to you, how can you put it right?

Although we have gone carefully over each variety of loudspeaker, explaining what each one does or should do, there are still a lot of things to keep track of. So here we want to pick up some of the things you might do wrong, or that might go wrong, so you know how to go about finding why things do not work out as expected, and can go about getting it right.

PLACEMENT PROBLEMS

If you are using one of the larger types of loudspeaker, that is a corner horn, a large infinite baffle, or a conventional bass reflex, wrong placement can result in loss of bass, making the reproduction sound thin as we have described. By bass, here, we do not necessarily mean the very lowest frequencies that the unit handles, or is supposed to handle, but those whose wavelengths are, say five or six times the dimensions of the loudspeaker.

What does that mean? Suppose your loudspeaker has an enclosure of some sort that is 2 feet wide by 3 feet high, and 18 inches deep. The average of those dimensions is about 2 feet, and five or six times 2 feet is 10 or 12 feet. So the frequencies that will suffer by wrong placement of this loudspeaker most are those whose wavelengths are in the region of 10 to 12 feet, that is, say 90 to 110 hertz.

This is of course relative. Above that frequency, the effect will diminish, and below it will get worse. And the frequency at which it becomes evident, which is what we quoted in the previous paragraph as between 90 and 110 hertz for the enclosure 2 feet by 3 feet by 18 inches, depends on the size of the box.

Not just one or two dimensions are involved here, but the average of the whole box. At low frequencies, where wavelengths are comparable with the dimensions we are discussing (which is why they become important there), the waves flow around surfaces, sort of fill the space available to them, as they move forward.

At higher frequencies, where wavelengths are shorter than the dimensions of the space in which they propagate, waves do not flow around objects in the same way. They move forward in bunches, thus try to maintain straight-line propagation. If an object of a few wavelengths in size gets in their way it produces a shadow behind it, because the waves do not flow around the object under these circumstances.

But at the frequencies we are discussing, relative to loudspeaker cabinet or box size, waves flow around; they do not necessarily stick to straight paths. So the average size of the object around which they flow is what determines the effect.

So, understanding that, it is important to put the box in such a position that it "persuades" the floppy sort of low-frequency wave out into the room as well as possible.

Have you chosen the best place for it? Take another look at Fig. 5-5. If a corner placement is not practical for whatever reason, you can improve the effect with a midwall placement by making the loudspeaker, or at least the part of it responsible for getting the bass out, face the wall. Stand it a few inches from the wall. If you play some program with good bass in it, preferably not just one note, you can adjust the spacing until you get the best bass result. (Fig. 9-1).

When you do this, you are, in effect, making the delivery into the room like an approximate horn mouth, like the folded horns do, but not so precisely. You expand the wave, before you let it go, and that increases the effectiveness with which the loudspeaker delivers waves of that frequency into the room.

Because you are now feeding the output from the cone into a space between the loudspeaker and the wall, this increases the air load on the loudspeaker cone, causing the space to absorb more sound energy from the cone. Then, as the wave spreads out in the space between the loudspeaker and the wall, the area it occupies expands from just the area of the cone up to the area represented by multiplying the perimeter of the loudspeaker box by the distance from the wall at which you eventually put it.

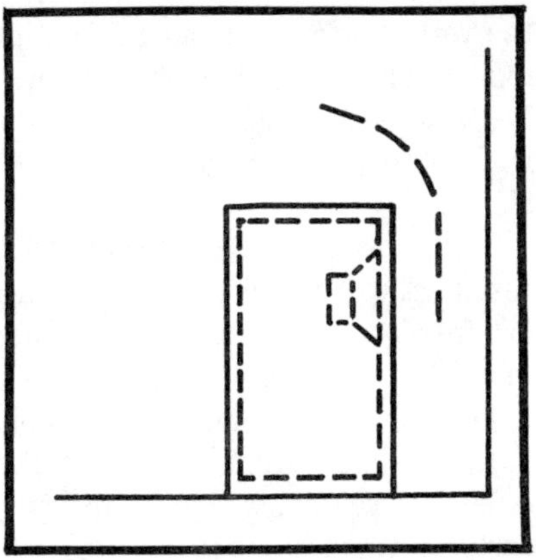

Fig. 9-1. A way to get better bass from a unit placed in a midwall position.

The effect is not unlike that of putting the loudspeaker unit into a baffle, although it looks quite different. What you are doing in each case is helping a smaller object to get out a big wave more effectively.

This applies mainly to the range of frequencies we mentioned, which depend on the loudspeaker size. The extremely low frequencies, down in the 20- to 30-hertz range, are much more dependent on what you close in the room—doors and windows—than they are on where you put the loudspeaker. You can regard doors and windows, or any other openings, as ports that let the extreme bass out of the room whatever kind of loudspeaker you may have used to get that bass in the first place.

IMPORTANCE OF SEALING

Progressing to the acoustic suspension type, the important thing to realize is that control of the diaphragm, or cone movement, is by means of air pressure in the back of the box. At the lower frequencies in the loudspeaker's range, the fluctuation of air pressure inside the box is considerable and would represent quite a large sound pressure if it were to escape into the room. So this is what it must not be allowed to do.

Of course, if you release the pressure from such a small box, into a big room, the pressure would no longer be there. But, if the box is complete, except for some imperfect seals, the big pressure inside will still be developed at low frequencies, near resonance. And because of the big pressure difference at this frequency between inside and outside, the leak will allow a rapid, if small, air flow.

This means that complete sealing of all the joints in an acoustic suspension box is vital to its proper performance. You are not likely to overlook a large opening that might be considered a port. Much more likely are little cracks that are not properly sealed. Through these, the fluctuations of air pressure inside will cause leaks resulting in alternate escape and ingress of air. Like any leak, it will cause a hiss.

Because the pressure is due to a low frequency, this means the result will be a modulated hiss. Its intensity will vary at the rate of the low frequency that causes it or, to be more precise, at double that rate. The precise tone of the hiss will depend on the size and shape of the crack through which the air leaks.

To some extent, the same thing can happen with a loaded reflex, which should also be thoroughly sealed, except where the ports or vents are supposed to be. At resonance, the pressure fluctuations inside a loaded reflex may not be very much less than those in an acoustic suspension, although the driven diaphragm excursion may be a lot smaller.

THE RIGHT UNIT FOR THE BOX

One thing that has sometimes caused trouble is using the wrong unit for the kind of box into which it is put, or making the wrong kind of box for the unit, which is the same thing in reverse. For example, some may think that, because the loaded reflex *splits the difference* on the efficiency question, they may improve their efficiency by converting an acoustic suspension unit, to loaded reflex.

What trying to make that change overlooks is the fact that the improvement achieved is partly because the loudspeaker unit itself is an in-between type, a compromise. A unit for use in a loaded reflex design has a diaphragm that is heavier than standard, but not as heavy as one designed for use in an acoustic suspension system.

The softening of the surround is not so severe either. Putting the two differences together, the unit for use in an acoustic suspension system may have an unmounted resonance of 2 or 3 hertz, very low. For use in a loaded reflex design, the resonance may be between 10 and 20 hertz, which is quite a difference.

At frequencies above the range where the special construction augments the bass by whichever method, the efficiency is mainly due to the relative weight of the cone, or diaphragm, and the air it has to drive. The heavier the cone, for its size, the less efficient the loudspeaker at these mid-range frequencies. Note that size enters it too because that controls the amount of air the cone moves to make a wave.

We are talking for the moment about midrange frequencies, those well above the range for which the unit is specially designed.

Now, if an acoustic suspension unit has an unmounted resonance of 2 hertz, and a unit for a loaded reflex has a resonance, unmounted, of 20 hertz, this could be because the acoustic suspension cone is a about ten times as heavy as the loaded reflex unit cone, and the suspension is ten times as floppy.

The change could be due to different proportions. But assuming that the cone of an acoustic suspension unit is ten times the weight of a similar sized unit for loaded reflex use, the loaded reflex loudspeaker will be ten times as efficient in the midrange as the acoustic suspension. Only the relative cone weights for their size matter here.

There is another difference. We have talked about cone movement. The acoustic suspension cone has to move much further at low frequencies than does the loaded reflex. Figure 8-7 shows the different construction needed for acoustic suspension to avoid distortion. This is necessary with the acoustic suspension for another reason.

The low-frequency unit must reproduce some frequencies up into the midrange. It is not practical to make a small unit like this just to handle, say, from 40 to 80 hertz. The crossover will want to be at least at 250 hertz and possibly higher. Now this means that when the unit is reproducing a frequency in the range between 40 and 80 hertz, for which the cone must move a long way, and is expected to reproduce more frequencies up to 250 hertz, those frequencies will be distorted too by intermodulation distortion unless the long coil is used.

With the normal short coil (Fig. 8-7), when it travels out of the gap either way, the drive force for the higher frequencies will be reduced. Thus the intensity of the midrange frequencies will be quite heavily modulated by where the cone is at the moment on its low-frequency wave.

Because the loaded reflex design reduces this low-frequency movement, it reduces this problem too. With both the acoustic suspension and the loaded reflex type unit, the enclosure moves the resonance up in frequency. But the change is much further for acoustic suspension than it is for the loaded reflex.

If the stiffness component, due to the enclosed air in a loaded reflex box, raises the resonance from 10 or 20 hertz to 30 or 40 hertz, there will be little movement at that frequency because most of the output will be due to the other part of the resonance: the air in the duct, or slave cone, resonating with the enclosed air. This is the change that pushes efficiency up for the loaded reflex in comparison with the acoustic suspension.

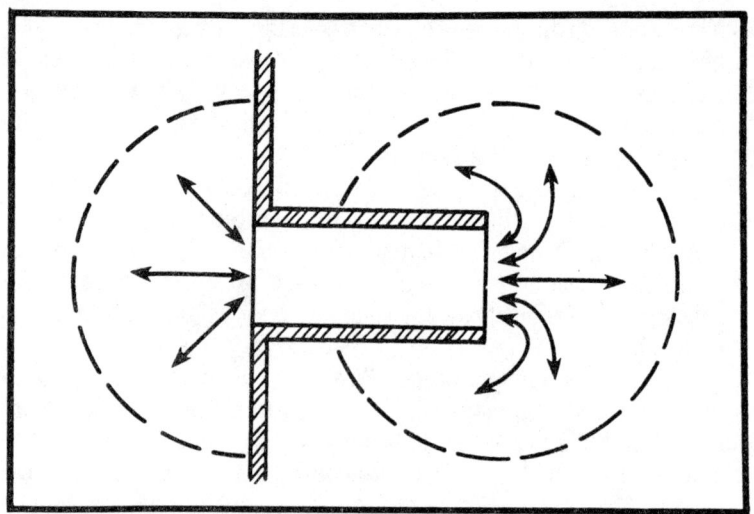

Fig. 9-2. Understanding the movement of air associated with a duct.

DUCT DESIGN

We ran through how to figure ducts in Chapter 8. That minimum equivalent length is based on the theory that both ends are open, sort of free and clear. The end that reaches the outside world should be flanged by terminating in the front panel of the loudspeaker cabinet. And the end that communicates with the air in the box should be completely open (Fig. 9-2).

The most common mistake is to overlook what happens at the inside end. There may be room physically to put in the duct, but you omit to notice that its exit is not facing open (but inside) air (Fig. 9-3). You can cure this, perhaps, by putting an elbow bend in it. And, if you want a really long duct, you can make it traverse back and forth to make a labyrinth.

The problem you have to watch then is how much the duct subtracts from the available volume inside the box. Do not forget that the working volume is what you have left, after the space taken by the duct is subtracted from it. If you find the duct using up more than half the available volume, take a good second look, you are probably making the duct take up too much of the available space.

If the duct is quite long, as it can be if the whole box is taken up with it, then what you are doing may be folding up the arrangement of Fig. 6-4. This would have to be at least a medium sized box to get some 5 to 8 feet of duct packed into it. But it is possible. In figuring out how it works, you need to think about this: is the air in the duct behaving just like a single mass at the frequencies where it comes into use, or is it behaving like a transmission line for acoustic propagation?

There are instances where it may be a bit of each. Then, if you figure it one way (and it would be extremely complicated to try to figure it out both ways at once), the result would be wrong because of the way it partially operates the other way.

Thus, if you figure length of duct to determine the effective mass for tuning, as we showed in the previous chapter, you are assuming that all that air moves simultaneously together. But if it is long enough to take an appreciable part of a cycle to travel from end to the other at the frequency on which you are figuring, your conclusion will be wrong.

The important thing is to be aware that ducts can have one of two main purposes and may occasionally do both together. They can be acoustic channels, that resonate by virtue of their length, irrespective of their cross-sectional area. They can serve as an acoustic mass, in which case cross-sectional area is very much a part of the calculation.

You can make the duct smaller, keeping its effective mass the same, by reducing its area and length in the same proportions so their ratio remains the same. However, in doing this, do not forget to calculate on the basis of equivalent length, rather than actual length, because just an open hole has an equivalent length, which must always be added to the physical length of the duct to get its equivalent length.

That is using a duct to load the reflex box. Often some kind of drone cone is used, usually in a somewhat larger hole than the equivalent duct would be. To increase mass in a duct, you increase length or reduce rate, or both. But when you put a cone in the opening, that is another way of increasing mass.

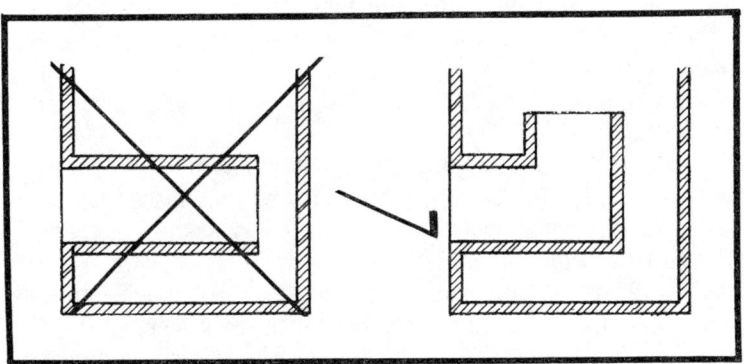

Fig. 9-3. The right and wrong way to put a duct in a reflex enclosure.

In fact, some of the newer loaded reflexes use a cone with quite a heavy weight attached to the cone. Doesn't this decrease their efficiency? To understand this, you need to remember the difference between the driven cone, which must also handle midrange, and this drone cone, which only comes into action near the resonant frequency of the system.

Thus the added weight helps get an even bigger mass swinging on the other end of the acoustic compliance, represented by the compressible air inside the box. It thus increases the energy taken from the primary, or drive cone. And it can transfer this energy to the air by the much bigger movement it thus generates, only in the vicinity of resonance.

Everything in this kind of design is, to some extent, a compromise. When you reduce the duct area to make it smaller, you are also reducing its coupling, both to the compressible air inside the box and to the outside air as a radiator. This makes the duct approach somewhat limited. Using a drone cone, with or without extra weighting, enables the same energy to be used for moving a larger surface area, thus getting more air movement out in the form of a sound wave for just the range near resonance.

Going back to the bass reflex, when the area of the port is about one-third the area of the cone or diaphragm, which means the port dimensions must be somewhat more than half the cone dimensions, the energy radiated by the cone and the port is about equal.

If you make the port smaller, it may still resonate at a frequency where it will augment response. But the energy is cut down in two ways: coupling is reduced with the air in the box and to the outside air as well.

But when you go to a duct, different relationships apply. Now the air in the duct has more mass relative to the outside air, and so it can radiate more energy for a given duct area than the simple port. The whole thing is a matter of compromise, and the number of possible combinations for a good design are very many. And the number of combinations that end up with a bad design are almost inevitably more.

Whatever is the optimum combination, it is a matter of putting the parts together in the right proportions so they make a proper match. But in considering this, you can make a mistake in calculating the actual duct length by taking that as the effective length, or the total box volume by taking that as the working volume, when these figures should be modified to find their effective values.

Thus you see you have to think of quite a lot of things in working with this type of loudspeaker, although in many respects it may have advantages over other types. If you have difficulty getting it straight, or if you have done something that does not work out, we suggest you read about these types carefully again to be sure you are thinking along the right track.

EFFECTIVE BOX SIZE

For example, if the total volume of the box is, say 1.5 cubic feet, then its working volume will be reduced by the space occupied by the vent or duct. If the duct takes up quarter of a cubic foot, you only have 1.25 cubic feet left in the box for expansion and compression. If the duct takes up ¾ of a cubic foot, the box has only the other ¾ cubic foot left, which can make quite a difference.

So optimum duct size must also take into account how much volume it robs from the box. You must also remember its physical need to have the inside end really open, as well as the outside end. If putting the cover on the box almost closes the inside end, it will not function as you expect, and you must modify it.

This is where using a drone cone can really help because it takes up no space inside as the duct must do. But the drone cone must also be right in several ways: as well as providing the correct match for the interior of the box and the drive cone that provides the pressure in there, it must resonate at the same frequency when the unit is all put together.

This is not easy to calculate. It is possible to do experimentally, but that too is not easy. If you plan this kind of system, we would recommend buying one ready-made, or buying the acoustic parts—the drive unit and the drone cone—with plans for the box in which to put it.

MATCHING THE ROOM

The earlier chapters considered matching the loudspeaker to the room in which it must work. The best idea is to pick units to suit the kind of room you have. But what if you have done it wrong, before you found out what suited?

Perhaps you can modify the loudspeaker you have or its placement, so it will sound better in that environment. The simplest modification may be to change the tone control settings on the amplifier a little. Doing this will not make the quality of sound approach what could be done, had you picked the right kind of system, but it may be possible to make the one you got more acceptable.

Then if you want to do more than that, without trading in loudspeakers to start over, you may consider taking steps to make

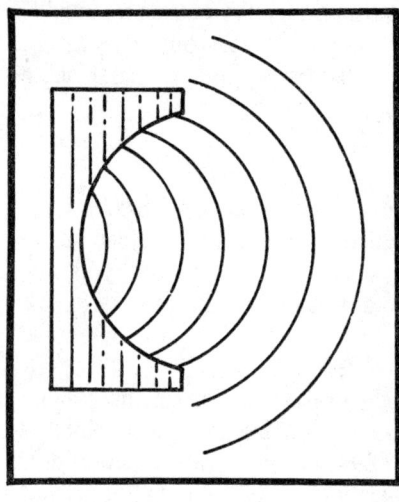

Fig. 9-4. How a conventional acoustic lens disperses a sound wave to simulate a spherical source.

the ones you have either more directional or more diffuse, according to which way they are wrong for your purpose. An acoustic lens is one way to do this.

Acoustic lenses have their effect by delaying sound passage through them. The zigzags, spaced at a distance smaller than the wavelength corresponding to the highest frequency, make the sound travel further than the direct distance, thus delay it.

The ones made by James B. Lansing Sound are designed to diffuse sound. They do this by producing more delay at the edges than at the middle (Fig. 9-4). Thus the middle part of the sound emerges earlier than the edges, producing a well-developed spherical wave, which is what the best kind of diffusing really does.

If you want a diffuse sound which your loudspeaker does not provide, that is fine. But you may also want to make a loudspeaker that diffuses quite well, more directional, which requires the opposite kind of lens. It must be designed the reverse way from the better known ones (Fig. 9-5). This delays a wave emerging from the center of the lens so it emerges more like a plane wave, thus is directional.

If you have access to a sheet-metal bender, or break, you can cut out sheets yourself. Then bend them alternate ways at about half-inch intervals, to get the zigzag effect. Then paint them, make some spacers (which will have to be cut on the slant), and assemble the whole thing with long threaded rods (Fig. 9-6).

That may seem like a lot of work, but it will do the job, if you exercise care to maintain unifomrity of the spacing of the zigzags. When it is finished and you have it working, you can show your

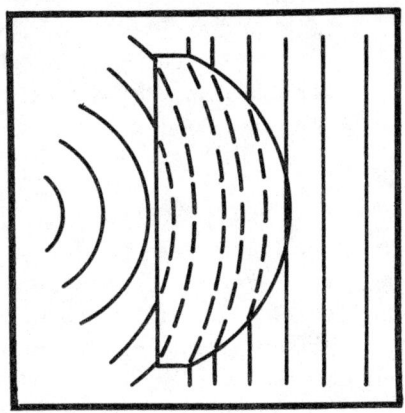

Fig. 9-5. A different shaping could be used to produce a beamed wave.

friends your own original design. Who knows, you might get into the business of making tailor-made acoustic lenses. Stranger things have happened.

If you have the idea what to do, it should enable you to tailor any loudspeaker system to suit any room within reason.

But in thinking of a loudspeaker as being directional or diffuse, remember that you need to be a little more precise than that. In being directional, are you thinking of being able to pinpoint sound? If so, under some circumstances, the property you want the loudspeaker to have is good diffusion, so that wherever you are listening you hear the sound directly from the loudspeaker.

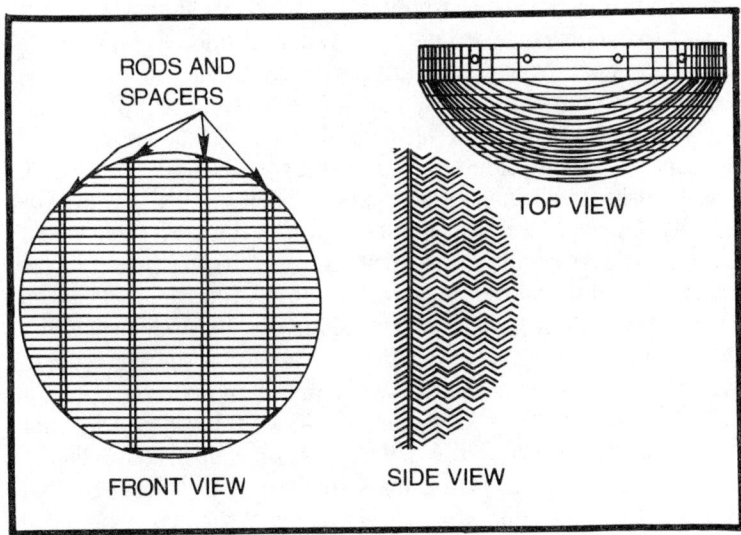

Fig. 9-6. Construction details for an acoustic lens to produce a beamed wave.

A directional loudspeaker will usually have the effect, at least in some positions, of meaning you do not hear the loudspeaker directly, only after reflection, which will destroy the pinpointing capability for which you are looking.

CARS

Next, let us take a look at installing loudspeakers in automobiles. Perhaps here the salvation is that most people like music in automobiles mainly for background, rather than for serious listening. Coupled with the fact that automobile interiors are well treated acoustically, simply to reduce exterior noise, this makes your problem easier.

The problem of placing car speakers is more mechanical than anything. Where can you put them that will harmonize with the interior decor of the automobile and not obstruct vision of the outside world necessary for driving?

We have already covered the possibilities of using oval units. If you are uncertain whether to use round or oval, it is probably better to use round ones for this application. In an automobile, increase of diaphragm size in one direction, such as from a 3-inch round to a 3- by 6-inch oval, will not effect much improvement.

If you are thinking the bigger the better, and you can find space for a 3- by 6-inch oval, consider whether you could manage to find space for a 4-inch round for example. Remember that the paneling in which you mount it will always serve as a baffle, so size is not all that important. In a car, especially with the windows shut, you do not have to move much air, to get quite a high sound level inside.

Of course, if you want to drive with the windows down so the world outside can hear your sound, you have a different problem—one we hesitate to tackle in this book. Because the sound, as heard outside, will be quite different from how it sounds inside, there is no way to make it uniformly good in both places.

Inside, you are effectively sitting inside the loudspeaker box, while outside the sound is what escapes from the same box. There is bound to be an acoustic filter characteristic between the two, so that the two sounds will be different whatever you do. Additionally, the difference will change, as you open the windows varying amounts, to change the acoustic coupling between inside and outside.

What we have been saying here applies to closed vehicles, not open ones, such as convertibles. The case for an open convertible is quite different again. Now you are sitting in a small "container" that has rather little effect on what you hear, except that it puts both you and the reproduced sound rather unnaturally close together.

Around you is all outdoors, going out in all directions for great distances. If you were driving along a shore line drive, and a ship out in the fog sounded its foghorn, you could probably tell the direction from sitting in a moving car, as well as you could if you were standing on the shoreline. Your ability to pinpoint direction comes from the fact that all outdoors is out there, without any reflections, at least in a direction that would confuse you.

Of course, if it is foggy where you are, as well as where the ship is, you would not likely be driving with the top down. But it is a hypothetical situation that shows the inconsistency of your listening capability in a convertible with the top down. Here you are, able to tell, at least roughly where the foghorn sound is coming from, and yet the sounds all around you from the loudspeakers mounted in the car are apt to sound confused because of the unnaturalness of the situation.

You probably even have sound program material that is designed for reproduction indoors, possibly over loudspeakers spaced as in an average living room. But in a closed car, a reasonable illusion would be created, because the car system is designed to do that. But in an open convertible, it becomes quite unnatural.

But we were discussing also, the effect of having windows open by different amounts in a car with the top covered. That is a different situation. It is something that is equally difficult, if not impossible, to do right. But if you close the windows you can get something acceptable.

However, if your object is mainly to let the outside world know you have sound inside your car, maybe what they hear is not too important to you. Perhaps you just want to make it loud. From our observations, those who have this objective are more concerned with having it sound loud than with having it be loud, which is not the same thing.

When a system delivers distorted sound, it always sounds louder than undistorted sound, even when the undistorted sound measures a much higher level on a sound level indicator. So we will not attempt to advise on this situation. We might find ourselves in trouble with the sound pollution authorities, for doing so.

SNUG MOUNTING

One thing we have not mentioned so far is the way a loudspeaker unit is mounted flush, which can make a difference to the sound that comes from it. We have mentioned sealing, and anytime you mount a loudspeaker unit in a hole be sure that the unit sets tightly into the hole so there are no leaks round the edges. If you install a piece of grill cloth, or an expanded metal equivalent, be sure that the diaphragm will not buzz against it, or that it will not buzz against some other part of the cabinet.

That can take careful examination to check. It may seem tight enough, but then you find a buzz, which may take a little tracking down. But that is something you can attend to with persistent attention.

A buzz is basically a mechanically produced sound. Sometimes an amplifier or other part of an electronic system can produce distortion that sounds just like a buzz. But if it sounds like a buzz, it most probably is a buzz. Only if you clear all possible sources of mechanical buzz, should you suspect the electronic parts of the system.

Your system produces acoustic waves that travel in air. You listen in air. So that makes spurious mechanical sounds, such as a buzz, difficult to track down. You tend to look at things you think might buzz, then concentrate to see if you can be sure. You may apply pressure with your fingers to see if you can stop or modify the buzz. But they can be very elusive.

A stethoscope can be quite a help for this kind of problem. It excludes the airborne sound, and enables you to listen to vibrations in the box or mountings, where the buzz may be originating. You can move the stethoscope around, noting whether response indicates you are getting warmer or cooler. Eventually you should be able to pinpoint the location of the buzz more easily this way, than by listening to sound in the air.

What we want to mention here is the general method of achieving good flush mounting. It is very easy, especially if you have a thick cabinet construction, to end up with the loudspeaker unit recessed considerably at the front. This is something to avoid, because that recess forms an acoustic cavity that will unnaturally color the reproduction from the loudspeaker.

Find a way to bring the loudspeaker forward so it is very close to being perfectly flush. There are various ways to do this. You could cut a few plies off the front side where the hole is, to allow the unit's frame to set in from the front perfectly flush. Or you could build up something with thin pieces of ply, or perhaps metal, to get a similar effect. The important thing, as well as getting a good flush finish, is to be sure that the unit mounts rigidly to the main body of the box. Also ensure that any piece of thin ply does not stand alone, where the sound can make it vibrate on its own.

10

The Importance Of A Thing Called Phase

You will not be around audio people for long before you hear them talk about phase, and how important it is or is not. So we decided to spend a chapter on it. If the word *phase* does not mean much to you right now, perhaps you are better off than those who think they know all about it—but don't. This is because what a lot of people *know* about phase happens to be completely wrong.

WHAT IS PHASE?

We have talked about frequency, about wavelength, and about time and the way waves travel. Phase links together all these facts in various different ways. The basic idea of phase derives from a theoretical way of generating sound waves, which is not altogether practical beyond the fact that it does provide us with a theoretical explanation of what we mean by phase.

Suppose a sound wave is generated mechanically by a small piston (Fig. 10-1) and the piston is driven by a little rotating shaft via a crank and connecting rod. Now we have a basis for thinking about phase. We start with the piston halfway up, as in compression.

Now, as the shaft rotates, 90 degrees will take the piston all the way up into compression, producing the pressure peak of a wave.

Another 90 degrees, making 180 degrees from the starting point, brings the piston back to the midway point again, except that now it is expanding the air in the wave, instead of compressing it.

A third 90 degrees, making 279 degrees rotation so far, takes the piston all the way to the bottom, producing the pressure minimum of the wave, or the maximum expansion.

145

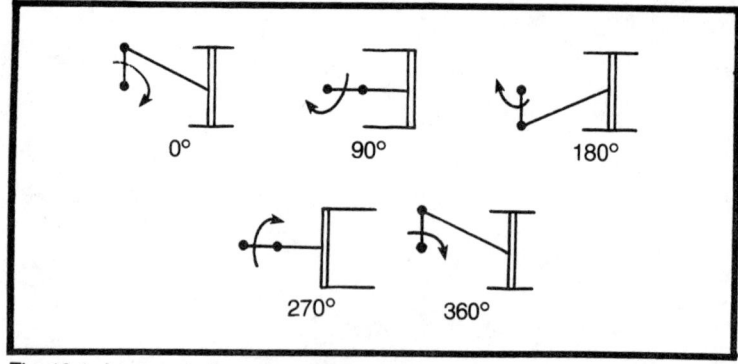

Fig. 10-1. A concept of wave generation on which the notion of phase angle is based.

Finally, the fourth 90 degrees, to complete the cycle of 360 degrees, brings, brings the piston back to its starting position again.

Each complete cycle of operation is called 360 degrees, and parts of a cycle are designated by the portion of a rotation that would be equivalent.

IS PHASE IMPORTANT?

Now the big question, about which you may hear arguments between audio people, is this: when you are listening to sound, which is transmitted to you by just such pressure fluctuations in a sequence that we call sound waves, can you tell differences in phase, or can't you?

People who argue make the mistake of thinking phase is just one thing, and that either it is important or it is not. As we shall see, there are ways in which it is crucially important, and ways in which, relatively, it does not matter at all. How can that be? If we try to answer the question academically, as so many of those who argue do, it can get confusing. It is better to work our way through various practical situations so we can see what happens in both theory and practice.

Let us start with simple sound waves being propagated across a room, space outdoors, or wherever. These simple sound waves could be just a single frequency, or a composite program of music or voice. They are a sequence of pressure fluctuations traveling across space that could be viewed as having been synthesized by as many little pistons as there are frequencies in the sound, each going up and down at its own speed.

The question is, to repeat it once again, can you tell differences in phase, or can't you? Would it make any difference to what you hear, or think you hear, if one of those pistons got, say a quarter of a revolution, or maybe a half of a revolution, out of its proper time?

So can you tell the difference due to such small changes in timing, or can't you? The immediate, first answer to this question is no you can't. Listening to a single tone, or even to a multiplicity of tones, all you can tell about them is the rate at which pressure keeps changing, known as its frequency. Their exact timing in this sequence of changes, is something to which our ears are completely insensitive.

Now, before you run off and say, "Okay, so phase does not make any difference," stay with us because that is not all there is to it by a long way. While that statement is true, perhaps the only way to grasp what phase really means, and when it is important, is to prove it for yourself because seeing—or in this case hearing—is believing.

First, let us be sure we know what we have said so far. We were talking about a simple sound that could be coming from one loudspeaker or from an actual original sound source. The experiments that show something different involve more than one loudspeaker that is supposed to be putting out the same sound.

You recognize that as applying to monophonic. So phase is important on monophonic, not on stereo? Hold on, we will get to that. Take it a step at a time. Phase can be important, even in one loudspeaker, particularly if it is a multiway type. So let us proceed with some experiments you can perform, or at least imagine performing.

Get two loudspeakers, just simple inexpensive ones will do in simple box cabinets to make them easy to set up, and connect them both to a single amplifier supplied with a mono program source of some kind (Fig. 10-2).

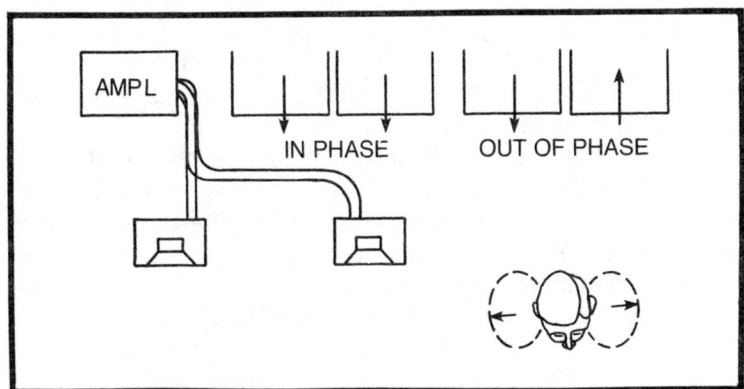

Fig. 10-2. A setup for checking the phasing of two loudspeakers.

CHECKING FOR PHASE

Place them, not too far apart, in a position where you can listen to both of them at once. Move your position, particularly your head from side to side relative to them, to see how the sound changes or how your perception of it does.

Two things are important here to get a good experiment. First, the loudspeakers should be identical in type and fed with sound at identical levels, such as by being connected in parallel to the same amplifier. Second, you need to listen at a position in front of them, where you are at an equal distance from each of them. This is the spot where your observations on phase will be most marked.

If your speakers are standardized, you can connect them so you know they are in phase by connecting the same terminals on each loudspeaker to each other, and to the amplifier output terminals. In some ways it may be more instructive to try a couple of loudspeakers that you do not know are standardized, because then you can use this method to find out if they are.

Play some good music that has a wide mixture of frequencies in it, including some good bass, through the two loudspeakers. As you listen, move your head from side to side, a foot or two, to see how the apparent source of sound shifts.

If the loudspeakers are connected in phase so both diaphragms move forward and backward together when your head is exactly in the middle, the sound will seem to come from a spot midway between the two of them. And as you move to one side or the other, the sound will seem to shift to the loudspeaker to which you come nearer (Fig. 10-3).

Only when you are exactly in the midway position will the sound seem to come from a point midway between the loudspeakers. As you move your head to either side, the apparent source of sound also moves toward the unit that you are getting closer to. For a range of positions near the middle, you will have an illusion that the sound is coming from somewhere between the loudspeakers rather than from one or the other.

But here is where it is important for you to use identical loudspeakers. If the loudspeakers have different frequency responses, then at one frequency the one at the left may be stronger while at another frequency the one at the right may be stronger. This will make different frequency components of the total sound seem to come from different positions between the loudspeakers.

However, even if the loudspeakers are different, and no two are completely identical even though they may be very close to it, when they are connected in phase the sound will seem to come from

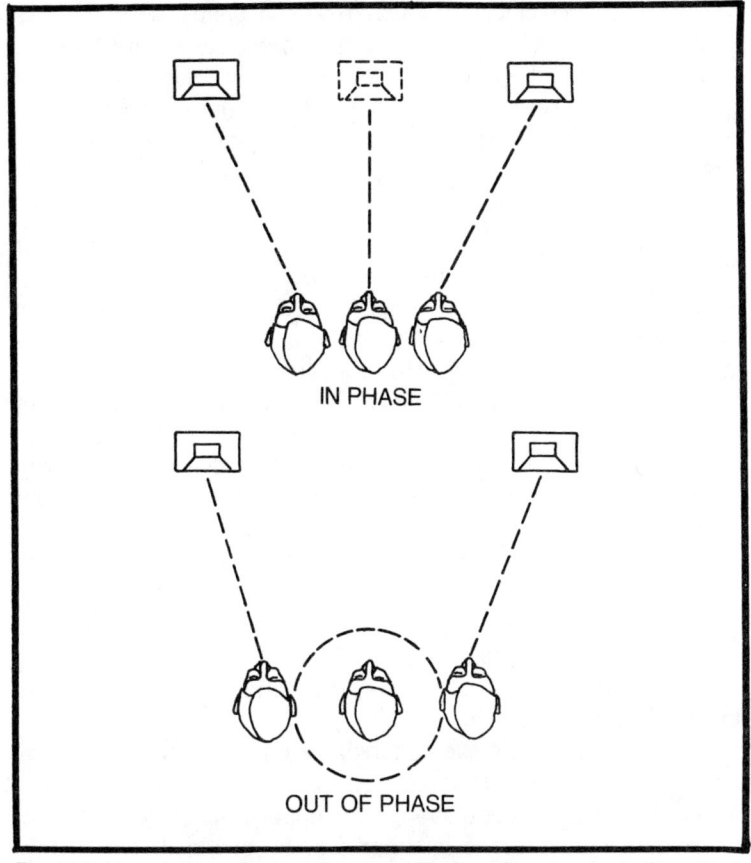

Fig. 10-3. How phasing changes the sound illusion when you are in the critical position, exactly midway between the units.

in between them. The more closely they are identical, the better will that in-between point be pinpointed. If they are not so nearly identical, the sound will come from a scattered, in-between area.

Now, reverse the connection to only one of the loudspeakers. If they were in phase before, they are now out of phase. This means that when one speaker's cone moves forward, the other one's moves backward, and vice versa. When this happens, the apparent source of sound no longer moves through a midway position.

In this condition, when your head is on either side of center, you can still identify the sound as coming from the speaker nearest to you. But as you move your head through the middle position, the sound becomes quite peculiar. It seems to lose all sense of direction, as if it is coming, not from any single place in the room, but is all round you.

149

Paying a little closer attention to what happens, when they are in phase as you move your head the sound moves its apparent source from one to the other; when you are exactly midway, so is the apparent source of sound. But when they are out of phase, as you move your head the sound leaves the nearer speaker on the opposite side from where the other one is, and when you are exactly midway, you get a maximum confusion of sound.

This is how you can know, quite definitely, the correct or incorrect phasing of two loudspeakers connected on the same program output: monophonic. We must emphasize the importance of connecting them monophonically for this test. If you put the speakers on stereo, it is not so easy to spot. You may notice a difference, but it will be almost impossible to be sure which is the correct connection.

When you have it right, the stereo will sound better, but finding out which way is right will prove difficult because you are not quite sure what you are listening for.

For example, if an instrument or an instrumental group is supposed to come from center stage, then out-of-phase connection will make them "come unglued," so they seem to come from all round the room. But that could also be how they are supposed to sound. You do not know. That is why monophonic connection is best for checking phasing.

The one thing that can be a help in a stereo situation, where both channels carry the whole frequency range, is the bass frequency response. In the midrange and upper frequencies, you may not know which effect is intended for the program. But at low frequencies, one connection will cause left and right to cancel one another, while the other will have them augment one another. The one where they augment should be the in-phase connection. This means you need to check it using program that has a good low-frequency bass content.

SYSTEM PHASING

You can often tell that something is wrong, or maybe get the impression that it is not as good as it should be. But pinpointing the cause is not easy. One example that well illustrates the problem was a long hall with a small gallery along the sides and back in which professional sound installers had placed four loudspeakers, two for the lower level, one on either side of the stage, and two upstairs for the galleries (Fig. 10-4).

Hearing in the hall proved difficult almost everywhere, except in positions where you were fairly close to one loudspeaker. We

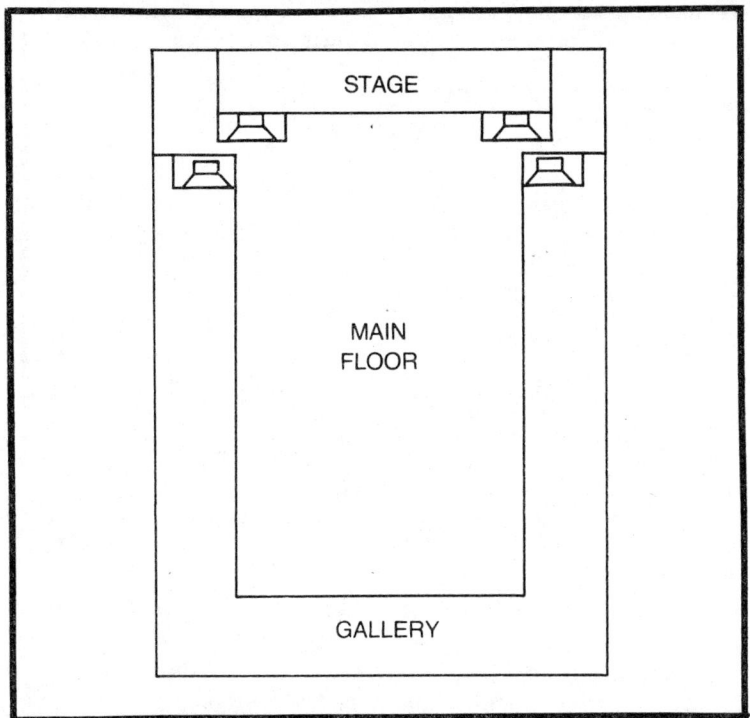

Fig. 10-4. A particular system where wrong phasing took some tracing down, which was instructive.

suspected phase of being the cause of the problem. But how do you find out which speaker, or speakers, is out of phase?

First, we disconnected the two gallery speakers and listened to only the downstairs pair (Fig. 10-5). That was okay everywhere. Then we connected the gallery pair and disconnected the downstairs pair. That was not so easy to check because sound was not bad along the side galleries with the downstairs units off, but in the middle at the back there was this out-of-phase effect.

Now, which one of the gallery speakers was wrongly connected? Had we only been using the two gallery speakers, it would not matter: reversing either of them would correct it. But they must also be right with the downstairs speakers, so we must find out which one needed reversing for that to be right.

When you have a situation like that, you review what you have. With all the loudspeakers connected the sound was confused, making speech rather unintelligible in a lot of places. In fact the only places where it was really clear was relatively near to any one of the speakers. Disconnecting the gallery speakers actually made the

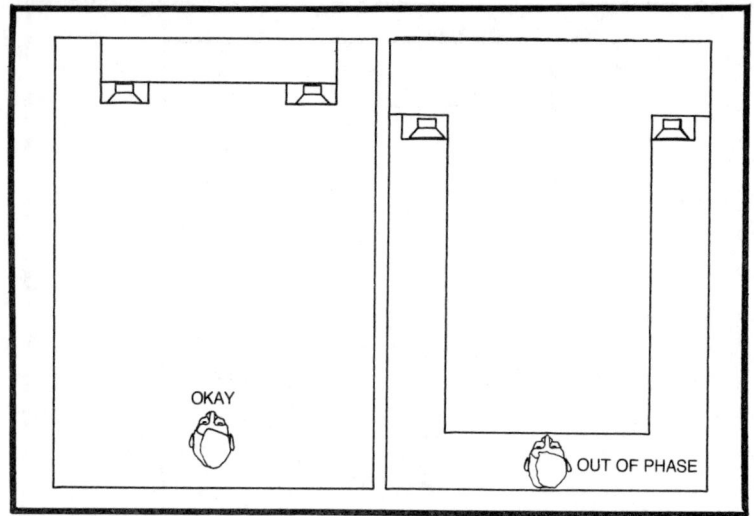

Fig. 10-5. Steps in the tracing of incorrect phasing for the problem of Fig. 10-4.

sound clearer everywhere, except close to the silent gallery speakers.

And with only the gallery speakers connected, those were the only places where you could hear anything intelligently.

To find out, we disconnected them the other way: by sides, rather than by up and down. We soon found out that with only the left side speakers on, the sound was much better than with only the right side speakers on, listening in corresponding spots. So we reversed the right side gallery speaker and connected them all back on. Now the sound was perfect everywhere.

Had that not worked, we had one more step we could have applied. We could have reversed either of the gallery speakers, then tried them each way as a pair, compared with the downstairs speakers as a pair. In fact that might be the better way to do it. In the case we were dealing with, the method of wiring made it not quite so easy to do. So, having satisfied ourselves by the method we did use, we left it at that.

The difference was convincing evidence of the importance of correct phasing. This also gives you an idea about how to proceed. Of course, that was a monophonic system, where the system with which you will more likely be concerned will be stereo or quadraphonic. But the best way to test any system for correct phasing is on monophonic.

Most stereo systems have a two-position switch that provides for stereo or mono. Or a quadraphonic system should have a choice of all three. Always switch to monophonic to perform your phase checks.

RELATIONS BETWEEN STEREO AND QUAD

Before the days of quadraphonic, a very similar effect was obtained, not quite so well, with a stereo system by mixing in program that is out of phase to the existing left and right channels. Let us see how that worked.

To get different positions in the apparent source of sound on stereo, the instruments have varying proportions of their individual sounds on left and right, always in phase, or very nearly so (Fig. 10-6).

Then to get an additional reverberant effect of sound coming from other parts of the room, a delayed sound was mixed into left and right, but with its two parts out of phase, which dissociates the sound from the loudspeakers (Fig. 10-7).

Remember how, when two loudspeakers are connected out of phase, the sound seems to come from all round the room, instead of up front? This is the effect that some stereo programs used, before the advent of quadraphonic to get ambience, or reverberant effects. First, the direct sound is given good position location in the normal stereo way. Then the delayed sound is added with phase reversed between the loudspeakers.

Note that phase relationship between the original and reverberant components is unimportant, because there is already a difference of several milliseconds in time between them, which will be a great many cycles of any of the component frequencies. The important thing for this method of reproduction, where the program was designed to produce the effect, was that the original sound should be correctly phased, and the reverberant components should be out of phase.

Stereophonic reproduction takes two channels of sound, one for left, one for right. Where tape recordings are used, those

ALL LEFT	MORE LEFT THAN RIGHT	LEFT AND RIGHT EQUAL	MORE RIGHT THAN LEFT	ALL RIGHT

Fig. 10-6. How stereophonic program varies the apparent position of various components along a line in front of the listener.

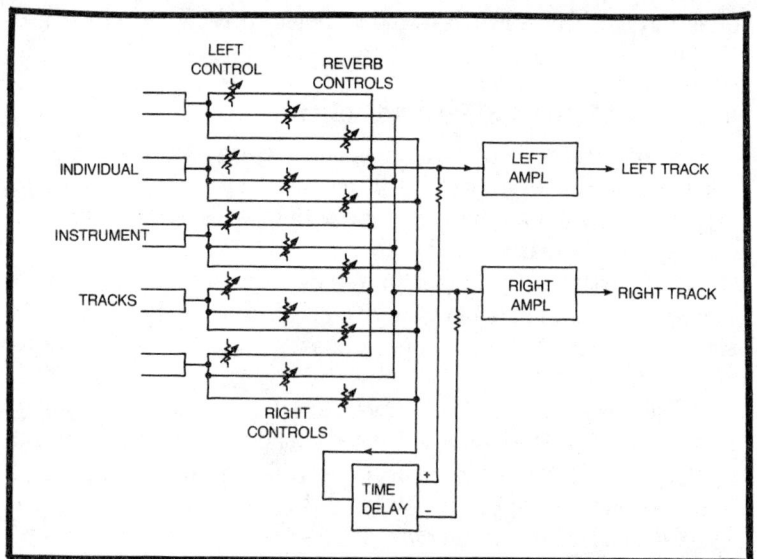

Fig. 10-7. A mixing system to give some sense of ambience from a stereo system, but not as precise as is possible with a quadraphonic system.

channels take two tracks. On stereo records, one channel takes each wall of the one groove, in which the stylus rides. On stereo FM multiplex, a different method of getting all the information into a single channel is used, so they can be separated out again at the receiving end.

The separation can use a variety of technical methods of which the best is probably switching. An ultrasonic switching frequency of 38 kilohertz is used to switch the audio output between left and right. A book on stereo multiplex will explain to you how this is done. The point we want to make here is that you have the equivalent of two channels available, whether you use tape recordings, stereo records, or FM multiplex.

To go quadraphonic, you should need four channels, instead of two, should you not? With tape, that is a possibility, although it would be more extravagant in the use of tape. With stereo records, you have only one groove, which has only two walls: where would you put the other two channels? And can you extend the switching technique on multiplex, to get four channels?

And as when stereo came in, a concern was to make program compatible with mono so it could be played on either mono or stereo equipment, and so equipment designed for stereo could play either mono or stereo. When quadraphonic came in, it should be compatible with both mono and stereo.

This fact led to the development of one accepted form of quadraphonic and others are similar. If an out-of-phase signal from the two channels produces an illusion of sound from behind—or as near behind as you can get from two speakers actually in front—while an in-phase signal produces a center-front effect, can't we use other phase relationships to portray sounds coming from other directions?

So really, the simulation or reverberant sound on a two-channel stereo system, where both loudspeakers ar in front of the listener, provided the basis for developing a way to put a sort of four-channel sound onto only two channels. With the stereo system, the effect was produced by using different phase realtionships on only two channels.

Quadraphonic does the same thing, with a little more precision. Some special electronic circuitry takes components of the composite left and right program signal that are already out of phase, and feeds them to the back loudspeakers, instead of to the front ones. Further refinements in the electronic digital logic circuitry enables the left and right speakers at the back to be fed with individual differences, similar to the differences that front speakers have always had on stereo.

The problem with trying to set a quadraphonic system up correctly if you are not already sure that your speakers are correctly phased, is that you cannot be quite sure what the program ought to sound like.

To illustrate the problem better, Fig. 10-8 shows where the sound will get routed according to the way the stylus would be moving on a stereo record recorded for quadraphonic reproduction. There are only four loudspeakers, or four channels (you could add extra loudspeakers if you want), but the sound can be made to come from virtually any angle round the room.

In addition to the movement that locates sound in each of the four corners, we have shown the stylus movement that will direct the sound to midfront, midback, and each midside.

Direction of stylus movement merely indicates the phase relationship between the signals originally known as left and right in a quite graphical way. The same relationships can exist, whether or not the program is transcribed on a stereo record, providing you are using two channels of some sort.

This means that the quadraphonic system has the capability of taking these two channels, processing them by means of digital computers so that the correct proportions of intensity are fed to four channels to duplicate the original inputs to those four channels. This is one of the finer achievements of modern technology. And it involves many phase problems of its own as you can probably imagine.

But that relates to phasing within the electronic part of the system. When you are installing loudspeakers, you are concerned only with phasing in what is now delivered to you again as four separate channels.

Now, to see what this means for phasing, suppose you have a quadraphonic system connected into an amplifier system that will provide four outputs, one for each speaker position. Now, each loudspeaker could be connected either way around. And reversing any one with respect to one other will change the result that those two of the four are supposed to produce.

To investigate the possibilities, any way you connect all four of them will give some form of quadraphonic illusion, but it may not be the right one. If you take one speaker output and change the other three, there are two possibilities for each of them, which adds up to a total of $2 \times 2 \times 2 = 8$ combinations, of which only one is correct.

Now, after you may have tried a few of the seven wrong ones, you will probably become thoroughly confused. By then it may even take careful listening to feel sure that something is wrong. This why it is best to check the system out with monophonic so you can tell much more easily which is right and wrong, step by step.

You phase the front two, one against the other, just as in stereo. You then phase the back two, one against the other. And finally, you phase the back two together against the front two together. Or if that gives you problems phase two on one side, with the other side disconnected. But if you do it that way, and find it necessary to reverse them (say you reverse the back one) then do not forget to reverse both back speakers, if you had already phased them against each other.

CROSSOVER PHASE

So much for attention to phase between different loudspeakers on whatever kind of system in the same room. Phase can also be important between different units in the same loudspeaker of a multiway system. For example, the midrange unit can be out of phase with the woofer, or the tweeter can be out of phase with the midrange, or some other combination may occur.

These possible differences are usually cared for in production loudspeakers by checking each type during production, and having a standard way of connecting them that is always right. But if you put

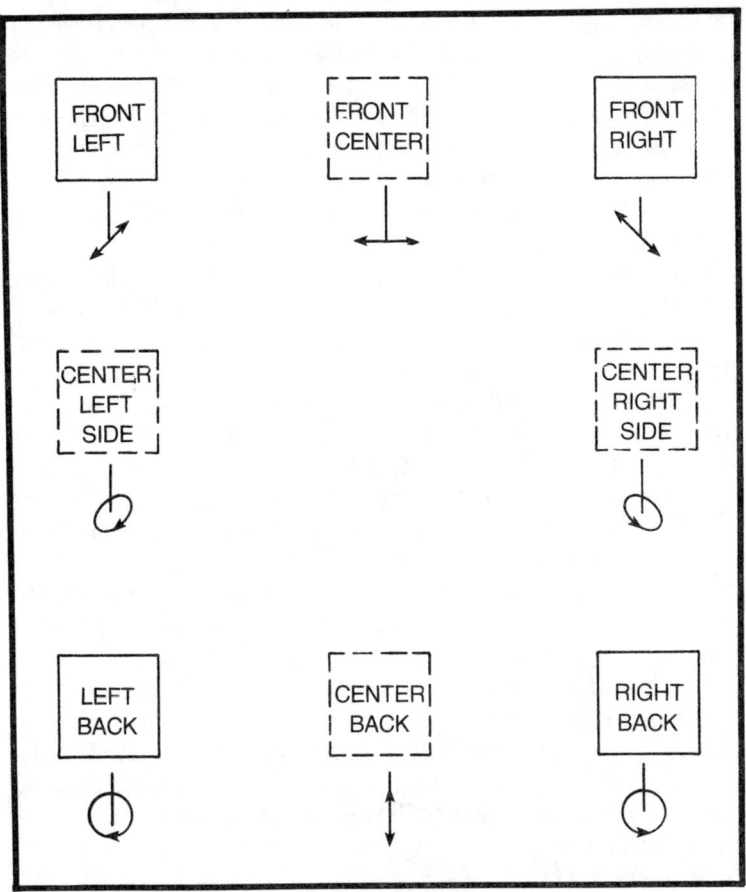

Fig. 10-8. Various stylus movements associated with a stereo record, recorded for quadraphonic reproduction with the apparent positions each is designed to reproduce.

together your own system, you could do it wrong unless it comes with directions that enable you to be sure it is right like production systems use.

What you need to know is what effect incorrect phasing has under these circumstances. At most frequencies the system handles you could probably not tell the difference. The problem occurs in the vicinity of what are called crossover frequencies, where both units are handling some of the total sound signal.

CHECKING CROSSOVER PHASE

If you have an audio oscillator handy, as some hobbyists do, the easiest way to tell is to sweep the oscillator through the whole audio range, paying particular attention in the regions where the crossover frequencies are. Do not listen to steady tones by allowing the oscillator to stay on one frequency because this produces standing waves in the room, which can fool you. Notice instead what happens while you are sweeping the oscillator's frequency fairly quickly.

In vicinity of crossover, if the phasing is wrong, you will lose the sense of direction about where the sound is coming from at just about the crossover frequency.

That method is fine, if you have an audio oscillator, which many of you will not have. So the only way to do it will be to listen carefully on program signal. In some respects this is a better way to do it, but you do need to listen quite critically to pick it up.

While using an audio oscillator enables you to listen to only one frequency more easily, using a composite program gives you the whole picture. Another problem with phasing, in any ordinary room environment, is the standing waves we referred to.

Have you ever tried to track down a constant humming sound by ear? If it is interrupted you can spot where it comes from fairly easily. But if it is constant, and you move your head around trying to pinpoint it, it gets louder and softer. And the apparent direction, if you can form an impression of it at all, keeps changing.

This is why you need something like an interrupted sound, or a sound that is changing in pitch, or just the constantly changing composition present in virtually any program material. While using the oscillator, taking care not to let the frequency stand still can help you concentrate on frequency. If you are already aware of frequency, at least approximately, you can probably do better listening to program material with those frequencies in.

Whichever way you use, do not have any other loudspeakers on in the room while you phase units of a multiway system. Any other sources of sound in the room will make it difficult to listen critically enough to the one you are checking.

Concentrate on components in the program sound that have component sounds in the vicinity of crossover frequency. Does the whole sound seem solid, as if it is all coming from one loudspeaker containing in its woofer, midrange, and tweeter? Or do sounds that come near crossover frequency seem to get spread around the room?

Where you notice a spread-around effect—that is at what frequency—should give you the clue to which crossover produces the out-of-phase effect. If you have a three-way system and the only out-of-phase effect occurs at the crossover between woofer and midrange, then reverse the woofer phasing and try again. If it is only at the crossover between midrange and tweeter, then reverse the tweeter. If it occurs at both crossover frequencies, then reverse the midrange.

HOW IMPORTANT IS IT?

If getting it right takes such critical listening with all the other loudspeakers disconnected—a condition in which you never listen—why is it so important? Will having it wrong make enough difference for you to notice at all when all of them are working? That is a good question.

The fact is, you may or may not be sure that something is wrong, when there is an error in phasing somewhere. But when you get it right, everything will sound better.

For example, if you have stereo, or quadraphonic, multiway systems and one of them has incorrect phasing between units, the stereo effect will very definitely be defective. If it is right between the units that are correctly phase, say the low frequencies, then it will be wrong where they are out of phase, which would be the midrange or tweeter frequencies.

But how are you to find that out? You could go around reversing things, with everything on at once until you drive yourself crazy, and still do not know for sure what is right or wrong.

To get the whole system right, you have to settle just one thing at a time, step by step, in ways that enable you to be sure each step is right. Then when you get the whole system correctly connected, you will notice considerable improvement, compared with what it sounded like when something was incorrectly phased somewhere.

In all of this work, you need to keep clearly in mind two objectives that can seem to be opposites. To illustrate this, back in the early days of stereo, some enthusiasts would put the woofer in one place, the midrange unit somewhere else, and a tweeter in yet a third place around the room. Played on orchestral music perhaps he could convince someone that was "as good as stereo."

But can you imagine what a piano solo would sound like? As if you were in the middle of it, with strings all around you. Or what about speech? A man's voice, put together from bits and pieces coming from all over the room, is not exactly natural. The object of good reproduction is to make everything sound like its real self.

Even in orchestral music reproduction, you want individual instruments to sound as if they are in one piece, not as if they float round the room, according to which note they happen to be playing at the moment. Or if an instrument plays a chord—more than one note at once—it could sound as if the notes from the same instrument come from different places in the room.

So while stereo, and even more so quadraphonic, should give you an illusion of sound that fills the room with a complexity of sound form, it should be a quite clear complexity that gives integrity to the various components. Remember brass instruments, strings, and woodwinds all play musical notes in the same scale with the same frequencies present.

There are not certain frequencies that are typically brass, strings, or woodwinds. Each note that each kind of instrument plays has a fundamental and harmonics that occupies a similar frequency range to those used by each other kind of instrument. You cannot separate them on a frequency basis. But each has its individual character, determined by the way the fundamental and overtones go together, build up, and decay.

So your system must be able to reproduce each of these things happening in different places, but quite solidly in each place. This is a pretty demanding requirement when you think about it.

Prior to multichannel reproduction—stereo and later quadraphonic—getting realistic reproduction of various kinds of program was difficult: if it sounded good on full orchestra, it was unreal on speech, and vice versa. One needed a large-sounding source, another needed a small-sounding source, while yet others were in-between.

Multichannel reproduction makes it possible to do everything with the same setup. But to sound good on all of them it must be done right. And phasing is a very important part of that.

While phasing can be quite unimportant in some respects, there are many instances, where it is very important to have it right.

11

Feeding Loudspeakers Correctly

The previous chapter covered just one angle of making correct connections to loudspeakers: making sure they are connected the right way around, or in the correct phase. But there are several other aspects of getting everything connected right.

If the amplifier or amplifiers match the loudspeaker or loudspeakers, the problem may be as simple as being sure you connect them the right way around, which is just a matter of phasing, but if you are figuring out your own installation, chances are there is more to it than that.

MATCHING

The novice usually has this situation rather unexpectedly: either he hears little or nothing at all, or what he does hear sounds peculiar. Occasionally, it will be too loud, and he finds himself unable to control it. Or it may be distorted badly.

All this comes under the heading of feeding your loudspeakers correctly, what the professionals call "matching." Anytime you connect one or more loudspeakers to an amplifier output, a number of conditions need to be met.

First of these is that impedance must match. Then you need to be in the right efficiency bracket, have somewhere near the proper damping, and, if you use more than one loudspeaker on the same amplifier output, have proper distribution of power between your loudspeakers.

Then, if you use multiway systems, you need the proper crossovers to ensure that the woofers get the low frequencies, the midrange the frequencies that should go to them, and the tweeters their range. That may seem like a lot to take care of, but taken in stages it is not as difficult as it sounds.

The first and most important step is to have impedance right, or somewhere near right. To make it seem easy, amplifier outputs and loudspeakers each have rated impedances. Thus ideally, for example, you connect an 8-ohm loudspeaker to the 8-ohm output of an amplifier.

When you can do that you can be sure it is right, but what if you have two or more loudspeaker units to connect to the same amplifier output? Do you connect all the 8-ohm units to the 8-ohm amplifier output? No, it is not that simple.

Anyway, even when you have the figures match perfectly, the actual impedance matching may not be as perfect as the matching numbers suggest. The fact is that a loudspeaker whose impedance is rated as 8 ohms may have an actual impedance that various, for example, between 5 ohms and 50 ohms at different frequencies within the range it handles.

An amplifier will deliver its rated output in watts into a pure resistance load of the named impedance, in this example 8 ohms. So it will not deliver the same power into either 5 ohms or 50 ohms. But you can usually be sure that it will deliver the best, least distorted output, into a loudspeaker rated as being an 8-ohm speaker, although at most frequencies the output will not really be the rated power.

Let us see what means using those figures. To pick some easy figures to follow, suppose the amplifier is set to deliver what rates as 18 watts into the 8-ohm load. The formula is $W = V^2/Z$, or turned around $V = \sqrt{WZ}$, where V is voltage in units of volts, W is power in watts, and Z is impedance in ohms. Using that formula V is the square root of 8 times 18, or 144, which makes 12 volts.

Now an amplifier is designed to deliver very close to a constant voltage at different frequencies, rather than constant power. So we are now talking about an amplifier with a 12-volt output, which happens to be 18 watts when the load is precisely 8 ohms. Now what is it when the load is, say, 5 ohms or 50 ohms?

At 5 ohms the power is 144/5, or 28.8 watts. At 50 ohms, the power is 144/50, or 2.88 watts. Quite a variation from the nominal 18 watts. But can the amplifier deliver the 28.8 watts if it is rated only at 18 watts? And will 2.88 watts sound as loud as 18 watts, each at their respective frequencies?

The lowest impedance a loudspeaker has, is usually at around 1000 hertz, where human hearing is near its most sensitive, so that it is unlikely that you would need 28.8 watts to give an output that sounds like 18 watts. In normal program material, the power level at these frequencies is lower than the maximum needed for lower frequencies.

Perhaps, to get the same loudness, 1 watt would do, or even half a watt. Assume it is half a watt. In 8 ohms, that would be 2 volts. Then 2 volts across 5 ohms would be 4/5 of a watt. So the amplifier will handle that.

Now suppose you matched the loudspeaker so the amplifier could give full power at 5 ohms. Then where the impedance is 50 ohms, it will give only one-tenth of full power instead of about one-sixth as when matched to 8 ohms.

Designing amplifiers and loudspeakers to go together is a compromise. Each has to guess a little at what the other does since they are usually designed quite separately. The object is to make the sound delivered as loud as possible for the power available, and with the least distortion all things considered.

At the frequency where the impedance is lowest, the loudspeaker sounds loud naturally with much less than full power. So designers bank on that fact to allow a bigger margin. At the frequency where the impedance is highest more power is needed, but you cannot sacrifice too much power at other frequencies just so you can get it all there.

When the impedance runs high, power is reduced without distortion because the impedance just does not take all the power the amplifier can give at that frequency. When the impedance runs low, the power the amplifier can give is reduced because to give it would cause a lot of distortion. But there the power is not usually needed, so we settle for less.

Now, whether the loudspeaker rated at 8 ohms has an impedance that ranges between the example we used, 5 ohms and 50 ohms, or between perhaps 3 ohms and 25 ohms, the case changes: we are using a different compromise. Unless we make detailed measurements of the actual loudspeaker unit in the enclosure or box in which you use it, we do not know what this range is. So we have to make the assumption that, if the loudspeaker is rated at 8 ohms, it will have an actual spread in impedance variaton such that 8 ohms is in there somewhere.

We tell you that, so you will realize that exact matching is not critical, because even if the figures match exactly, an exact match does not exist. However, in what follows, we do base calculations on these rated impedances, which will divide the nominal power between a number of speakers connected to the same output. To refer to such relative power that each receives, we call it *nominal power*.

Where you cannect just one loudspeaker to each amplifier output, you really do not have to bother with this. Your loudspeaker has an impedance rated at some figure, such as 1.35 ohms, 4 ohms, 8 ohms, or 16 ohms, and the amplifier has an output with matching numbers. You just connect your loudspeaker to the terminals with the correct numbers.

MULTISPEAKER SYSTEMS

The problem comes where you connect more than one loudspeaker unit (and here we are not talking abut multi*way* systems, which we will come to later in this chapter) to the same amplifier output. For example, if you build your own directional array using a number of inexpensive units, you will have to do some impedance figuring.

We could give you all the formulas first, then show you how to apply them. But formulas are rather indigestible things, so we will do some figuring first to show you how it goes, then give you the rules so you can go on doing it for yourself.

The best way to get into this is to look at the various possibilities. Suppose you have a column consisting of six identical units, each rated as being 8 ohms (Fig. 11-1). If you connect them all in parallel, the combined impedance will be found by dividing the 8 ohm figure by 6, giving 1.33 ohms.

If your amplifier has a 1.35-ohm output, this will work fine. But suppose it does not, what other possibilities do you have?

If you connect three of them in parallel, then connect the two sets of three in series, each three in parallel will have an impedance found by dividing 8 ohms by 3, giving 2.67 ohms, and the two sets in series will be twice that, or 5.33 ohms. Probably the 4-ohm amplifier output will be close enough to use for feeding that arrangement.

If you connect pairs of units in parallel, then put the three pairs in series, each pair will have an impedance of half 8 ohms, which is 4 ohms, and the 3 pairs in series will be 3 times 4 ohms, which makes 12 ohms. You could try either the 8-ohm or the 16-ohm amplifier output to see which performs best with this arrangement. However, when you change amplifier connections, adjust the gain because just making that change may change either the level, or the point at which distortion begins.

If a single 8-ohm loudspeaker has values that vary between 5 and 50 ohms for different frequencies, then the combination of them

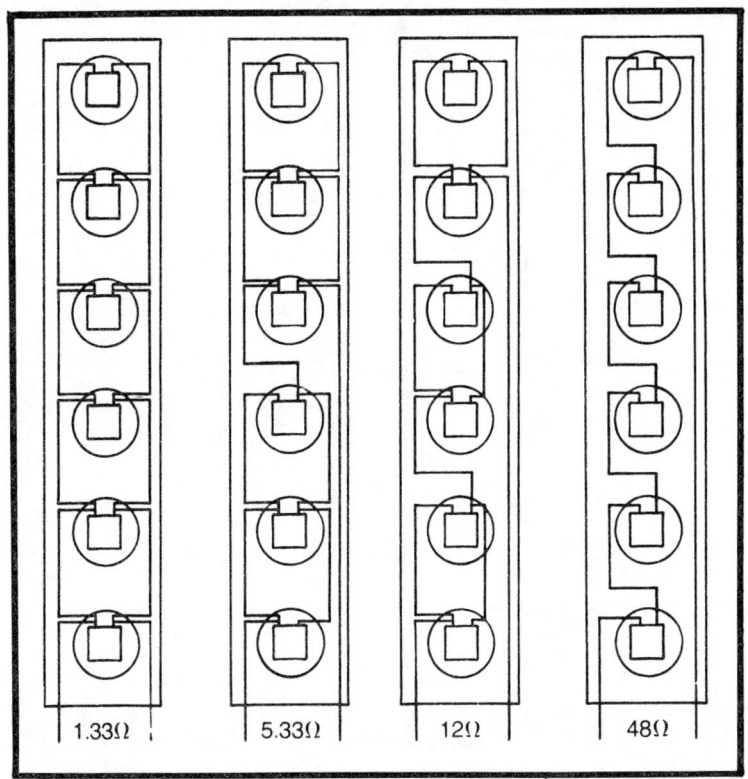

Fig. 11-1. Various ways of connecting a column of six indentical units, so each receives equal power, to show the impedance possibilities.

that makes a nominal 12 ohms will vary between 7.5 and 75 ohms. That sounds a little high for the 8 ohm output. Probably the 16 ohm output would be better. But the only way to be sure is to give it a try each way.

As well as just changing the tapping to which you connect the load, you want to adjust the gain setting. Connecting a load that varies between 7.5 and 75 ohms to the 8-ohm load will probably allow you to turn the input gain up until it gives rated output at the 7.5-ohm frequency. That should be pretty loud.

But now, changing to the 16-ohm output, still with a load that varies between 7.5 ohms and 75 ohms, would overload at the same gain setting. It would also sound louder because the voltage would go up. So to get a fair comparison, you need to turn the gain down until the distortion stops.

Note the setting of the gain control for best output level with each output connection. Then make the change as quickly as possi-

165

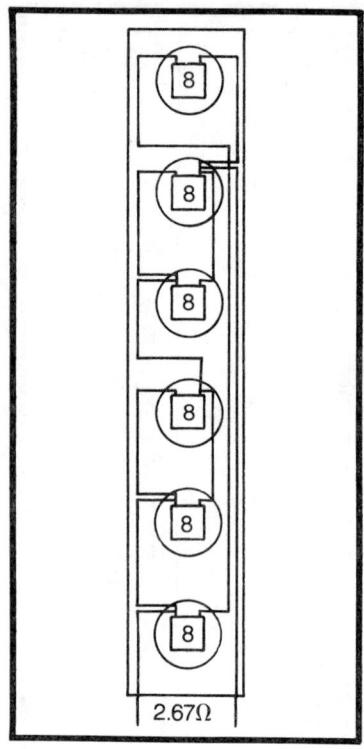

Fig. 11-2. One possibility of making connections that drives end units harder than the middle ones. This one makes the difference too marked.

ble, changing the gain setting at the same time, to get a direct comparison to find out which is best.

Finally, if you put them all in series, the arrangement will have an impedance of 6 times 8 ohms, or 48 ohms. If your amplifier should happen to have a 50-ohm output, which is not usual for a high fidelity or stereo amplifier but may come on a public address amplifier, that would do. We will have more to say about this choice later in the chapter, under the heading of *Damping*.

Now, in all of those arrangements each of the units gets the same nominal power as all the rest. If the total rated power is 30 watts, then each unit would get a nominal power of 5 watts. That is a satisfactory way to operate a column set of units, but it is also good to give the end ones a little more power than the in-between ones. However, it should not be more than double, preferably less.

You can make such a difference with six units, but the difference is too radical (Fig. 11-2). If you connect the four middle units in two pairs, then each parallel pair will be 4 ohms, and the two pairs in series will make 8 ohms. Now, if you connect the end ones directly in parallel with this, the two of them will make 4 ohms which, in parallel with this 8 ohms, brings the combined impedance down to 2.67 ohms. We will go into how to figure that in a minute.

Here the important thing is to calculate the relative power the various units receive. If the total rated output is 30 watts, you have, in effect, three 8-ohm impedances in parallel. So each will get 10 watts. That is 10 watts each for the end units, and 10 watts for the middle ones altogether, dividing down to 2.5 watts apiece. That is too big a difference in power.

It always helps to figure in volts as well. Therefore, 30 watts into 2.67 ohms would be the square root of 80 volts, which is nearly 9 volts. Across each of the end 8-ohm units, 9 volts will produce 81 divided by 8, which is just over 10 watts. And the middle group of four will divide the 9 volts into two 4½-volt pieces, each across two 8-ohm units. Hence, 4½ volts into 8 ohms gives 4½ squared, which is 20.25, divided by 8, makes just over 2½ watts, as we found the other way.

We can come up with some better arrangements using five units instead of six. If you put the middle three units in series (Fig. 11-3), that makes 3 times 8 ohms, or 24 ohms. Then the end two in series makes twice 8 ohms, or 16 ohms. Then 24 ohms in parallel

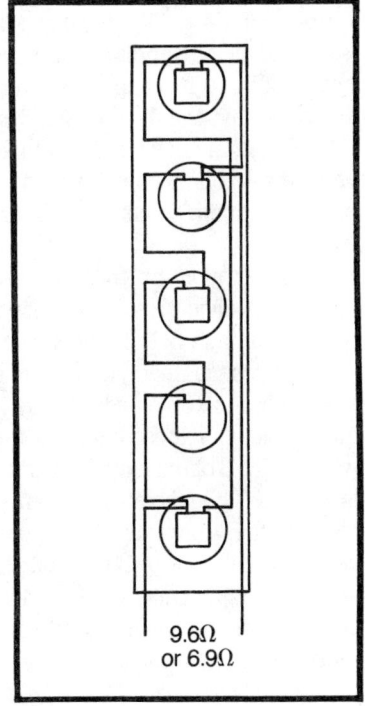

Fig. 11-3. A possibility, using a 5-unit stack with either identical units or units having different impedances.

with 16 ohms gives an impedance of 9.6 ohms, for which the 8-ohm amplifier output is probably close enough.

But how does the power distribute? Of 30 watts rated power, the 16 ohms will get three-fifths and the 24 ohms will get two-fifths. That is 18 watts and 12 watts, respectively. The 18 watts serves the two end units, giving them 9 watts apiece, and the 12 watts serves the middle three, giving them 4 watts apiece. While this is not such a radical difference, it is still greater than we want.

Or doing it by volts: take an easy figure, that can divided by both 2 and 3, such as 12 volts. The end units will get 6 volts each, the middle units 4 volts each. Hence, 6 volts across 8 ohms gives 36 divided by 8, or 4½ watts. And 4 volts across 8 ohms gives 16 divided by 8, or 2 watts. That is the same ratio as the 9 watts and 4 watts we calculated the other way.

You could improve this by using 4-ohm units in the middle, and 8-ohm units on the ends. The middle units now add up to 12 ohms, and the end ones still to 16 ohms. This means 12 ohms in parallel with 16 ohms is a combined impedance of just under 7 ohms, which is probably still okay to connect to the 8-ohm amplifier output.

But here we are more concerned with finding the power distribution. We can make the calculation easier by using a nominal 28 watts, rather than 30, because it divides by 7. Additionally, 12 and 16 is 28, and the power distribution between parallel groups is inversely proportional to the impedance numbers. Thus the 12 ohms will get 16 watts, and the 16 ohms will get 12 watts.

The 12 ohms consists of three units, so each gets 5 1/3 watts, while the 16 ohms consists of two units, so each gets 6 watts. That may make them so close to the same, that you cannot tell they are different.

Again, by volts: the end units still get 4½ watts. The middle ones get 16 divided by 4, which is 4 watts. This is in the same ratio as the 6 watts and 5 1/3 watts calculated the other way.

Suppose instead you use all units in parallel, but make the end ones 8 ohms and the middle ones 16 ohms each (Fig. 11-4). Now the middle ones have a combined impedance of one-third of 16 ohms, which is 5 1/3 ohms, and the end ones have one-half of 8 ohms, which is 4 ohms. The whole thing comes to 2.28 ohms, which might work okay on the 4-ohm tap.

But now, out of 28 watts, the 8-ohm units each take 8 watts, and the 16-ohm units each take 4 watts. In this case, we could get the same figures by assuming the amplifier output is 8 volts. Into the 8-ohm units, this will give 8 watts. Into the 16-ohm units, it will give 4 watts. Those are some of your choices. But when you use units

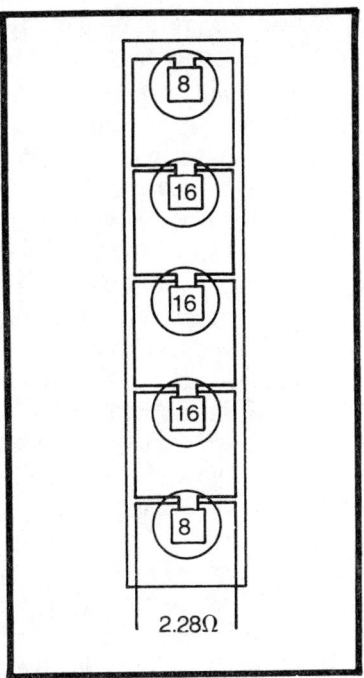

Fig. 11-4. Getting different power by using units of different impedances.

having different impedances, you have no assurance that their efficiencies are the same as they can be expected to be with identical units, so nominal electrical watts may not be a reliable indication of relative acoustic watts. You will have to try them out together to find what you have really got.

FINER ADJUSTMENT

Perhaps a better way that would work equally well with six units, or for that matter with any number, would be to use a resistance, or some resistances, to adjust the relative power more finely (Fig. 11-5). Thus, as we said at first, six 8-ohm units directly in parallel would make 1.33 ohms. But if you put the end two straight in parallel, that makes 4 ohms. Then, after paralleling the remaining four units, which makes 2 ohms, you put a 1-ohm resistor in series with them, making 3 ohms. The combined impedance, 4 ohms and 3 ohms, makes about 1.7 ohms.

THE FIGURING

If you have been following along not quite certain how we did the figuring, it is getting complicated enough now that we should give

169

you either the rules or the formulas. We find the rules easier to handle than the formulas and expect you will too. So here they are:
1. When you parallel a number of identical units, the combined impedance is the single impedance of each, divided by the number of units in parallel.
2. When you connect a number of identical units in series, the combined impedance is the single impedance of each, multiplied by the number of unite in series.
3. When you connect different impedances in series, the total impedance is found by adding them together. And the total power divides between them in proportion to their respective impedances. Thus 3 ohms in series with 4 ohms makes 7 ohms. And 28 watts would divide so that the 3 ohms gets 12 watts and the 4 ohms gets 16 watts.
4. When you connect different imedpances in parallel, the combined impedance is found by multiplying two of them together, then dividing by *their* sum. Using the result from the first two, go onto the third and repeat the procedure. Continue the product-over-the-sum procedure until the final impedance is calculated. And the total power divides between them in inverse proportion to their respective impedances. Thus 3 ohms in parallel with 4 ohms makes an impedance calculated as 4 times 3 divided by 4 plus 3, or 12 divided by 7, which is about 1.7 ohms. And a total power of 28 watts would go, 16 watts to the 3 ohms and 12 watts to the 4 ohms.

Applying that to the calculation we started a little way back, the end units would get half of 12 watts, or 6 watts each. The whole middle group, including the 1-ohm resistor, would get 16 watts. But the resistor takes one-third of that, or 5 1/3 watts, leaving 10 2/3 watts for the speakers. The individual middle speakers get one-fourth of that, or 2 2/3 watts each.

Perhaps here there is an even bigger advantage to working in volts than in just watts. Assume the amplifier output is 9 volts. Then the end units get 81 divided by 8, or 10⅛ watts. And the middle units get 6 volts, as the 9 volts divides between the resistor and the units, so their share of the wattage each is 36 divided by 8, or 4½ watts. You will find that 10⅛ and 4½ are in the same ratio as 6 and 2 2/3. Both are a ratio of 2.25 to 1.

This is more than a 2:1 ratio, so let us see what using a resistor of 0.5 ohms would do. Now the middle group adds 0.5 ohm to the 2 ohms for the speakers, making 2.5 ohms. The parallel combination

Fig. 11-5. Getting finer control of power to the individual units by use of a resistance.

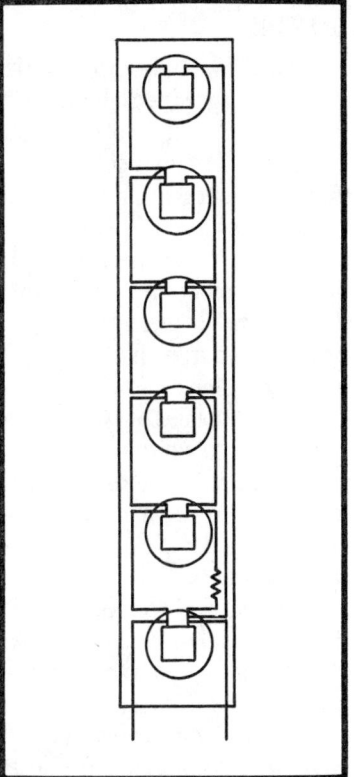

of 4 ohms and 2.5 ohms, is 4 multipled by 2.5, divided by 6.5, which figures to about 1.5 ohms.

Assuming we will find some way to match, let us look at the power division, and calculate what that is. To make the calculation easy, use a nomimal value that is a multiple of 6.5, such as 26 watts. Now, the middle group, including the resistor, get 16 watts and the end ones get 10 watts, which is 5 watts each. Of the 16 watts in the middle group, the resistor gets one-fifth, which is 3.2 watts, leaving 12.8 watts for the speakers, which is 3.2 watts each.

Doing that by volts, pick 10 as the amplifier output. The end units get the full 10 volts, which gives 100 divided by 8, or 12½ watts each. The middle units get 8 volts, which is 8 watts each. Now the ratio between 12½ and 8 is the same as the between 5 and 3.2. Each is 1.5625 to 1, or 25:16.

That is probably about right. You will need a 0.5-ohm resistor, capable of dissipating 3.2 watts or so. If you have a 5-watt rated 0.5-ohm wire-wound resistor, that would do fine.

DAMPING

So far, we have mixed series and parallel connection of speaker units just to get the right proportions of power, without considering whether series or parallel is better when you have a choice. As far as power distribution is concerned, there is no difference if the proportions work out the same. What does make a difference is something called damping.

When a loudspeaker is driven with an audio signal from an amplifier, sometimes it will overshoot, making sounds that were not due to signal delivered to it by the amplifier. Damping is intended to prevent this from happening. Amplifiers provide what is called a *damping factor* to do this.

If an amplifier output is rated at 8 ohms, meaning it needs a load of 8 ohms into which to deliver its rated power, then its own internal impedance will not be 8 ohms. Usually, the amplifier's internal impedance will be a much smaller value, such 0.8 ohm. That would represent a damping factor of 10. If it was 0.4 ohm, the damping factor would be 20.

When used to feed just one loudspeaker, that internal impedance of 0.8 ohm, or 0.4 ohm allows currents to circulate in the voice coil to stop any overshoot movement short, like putting brakes on the diaphragm.

Let us look at how that works in voltage effects. Suppose we have 10 volts output into an 8-ohm loudspeaker (which, for the purposes of illustration, we assume to be 8 ohms) with this damping factor of 20. The movement of the voice coil produces a voltage to match the 10 volts output. For the moment, we will ignore the voice coil resistance, which might mean that only half that or less is needed to match the amplifier's voltage.

Now, if that movement continues, after the amplifier's drive is removed, the 10 volts will still be generated by the moving voice coil, or perhaps something less than 5 volts if we take a more realistic figure. But, if the internal resistance of the amplifier is 0.4 ohms, that will load this voltage down to less than a volt. The current flowing in the voice coil, due to this loading, will virtually stop that movement in its tracks.

Try to get a picture of what happens. When the amplifier is delivering its output voltage, this drives the voice coil to a movement corresponding with the voltage applied. When the voltage stops, if the voice coil was disconnected, or connected to an amplifier with a higher output internal resistance, the movement could go on, making sound that the amplifier output did not "authorize."

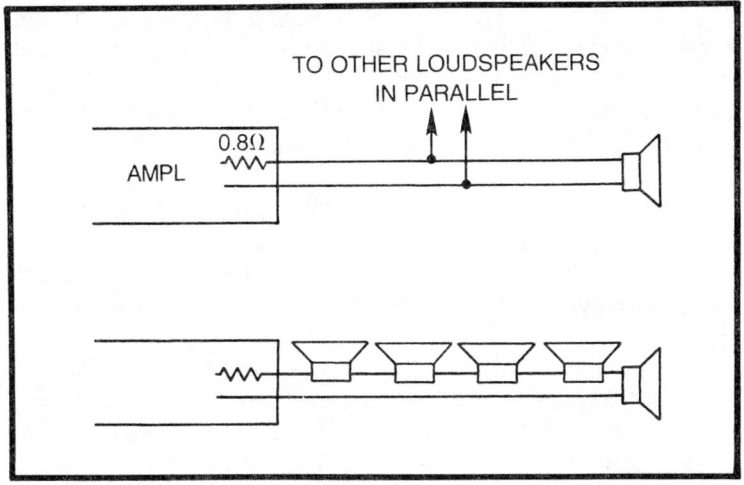

Fig. 11-6. Effect of series of parallel connection on damping of individual loudspeaker units.

But by having a low resistance, continuance of the movement associated with the voltage just discontinued, causes a current that stops the movement also. The lower the resistance, the better it stops it. Hence the term damping factor, which in a higher number, the lower the resistance. If the internal resistance is one-twentieth of the nominal load resistance, which is the example we took, then the damping factor is 20. The reciprocal is used.

This tells how amplifier damping affects a single loudspeaker connected to an amplifier output. But what about when a number of units are connected to the same amplifier output? How does this affect damping factor, or what the damping factor does? This depends on how the number of units are connected.

Now, if you connect loudspeaker units in parallel (Fig. 11-6) the amplifier's internal impedance is still connected directly across each loudspeaker unit's terminals, so that every one of them has this damping available to it.

But, if you connect them in series, then each unit has however many other units are in series with it, before the low internal impedance of the amplifier can act on it. Only if all the units happen to require identical damping—that is, if their tendency to overshoot is identical—will the amplifier's internal impedance put the brakes uniformly on all their diaphragms. Otherwise, what happens is that a lot of *pushing and shoving* occurs between the various units, leading to quite an erratic effective frequency response, especially to transients that produce over-shoot effects.

Most loudspeakers have main resonance frequencies that are slightly different individually, as well as a number of minor resonant frequencies that also differ from unit to unit. And there are antiresonant frequencies, quite often close to resonant frequencies. One produces a high impedance, the other a low impedance, at the loudspeaker terminals.

When a number of loudspeaker units are connected in series, the different variations in impedance, due to these differences in acoustic characteristic, combine in a way that results in very uneven voltage distribution between them at certain frequencies. That is not all.

Suppose you have just two loudspeakers in series that are moving differently but only slightly so when a drive voltage stops. Now the low internal resistance of the amplifier is in series to the two units, so that their voice coils are effectively in parallel. A voltage that either one generates, due to its remaining movement, will be fed to the other one, through the low internal resistance of the amplifier. This means that movement of one cone will drive the other, or vice versa.

This is what we meant just now by the pushing and shoving that occurs between units. If you have more than two units in series, then the combination of spurious movements becomes even more complex. The point is that, when they are all connected in parallel, the low internal resistance of the amplifier output is connected directly across each and every voice coil, so they all get stopped in their movement by the amplifier's damping factor.

The effect of a good damping factor is a sort of crisp, clear-cut sound. If you take a single loudspeaker unit, unmounted and unconnected, you will find the cone has a certain freedom of movement and, if you tap it, you will hear a short plop that emphasizes its resonant frequency, very much like the effect produced when you throw rocks in deep water.

Now, if you short-circuit the voice coil of this unit, and tap it again, the response will be quite different. If you try just pushing the coil, with and without the coil short-circuited, it will offer noticeably more resistance to movement than with it open. The effect of the short-circuit, or of a low resistance across the movement, is to put the brakes on.

Thus damping has the effect of stopping movement immediately after the driving force comes to a halt.

What this says is that parallel connection is to be preferred, whenever there is a choice, as providing better damping to individual units, and minimizing interaction between them.

HOW MUCH POWER?

We have talked about how the power we have will divide, but we will always want to know how much power we need. One way would be to take the loudspeaker ratings, if there are several of them connected together, multiply by how many, then get an amplifier that will deliver that much. But you may not need that much.

The speaker units you use in a column assembly, for example, may be rated at 20 watts each. So if you use 6 of them, your amplifier should provide 6 times 20 watts, or 120 watts output on each channel, according to that method of calculation.

Why you might not need that much is that using speakers in this way puts the sound energy where you want it, not all over the place. As a result, much less than 120 watts may be perfectly adequate because you will the speakers so they are "loafing."

On the other hand if each unit is rated at 20 watts, working by itself, when you put it in a column, it will be able to handle more because the other units work with it to hold down movement. It is not unusual for a column of six units rated at 20 watts each to be operated at a 200-watt level quite cleanly, for example, in a rock concert.

One problem with hearing is being able to hear what you want to hear, above what you don't want to hear. In some kinds of room, this means being able to hear the direct sound above the reverberant repetitions of itself. Reverberant repetitions can become so confusing that you cannot sort anything out.

Column speakers are one way of help with this problem. But regardless of the method, here is an important thing to note: when someone complains that they cannot hear, you need to find out why. If there is extraneous background noise coming from outside the room or building, or perhaps from air conditioning equipment or the like inside it, what the wanted sound must do is override that background sound so you can hear.

In that circumstance, whether the air conditioning is on or off, or whether whatever other extraneous sound is present or absent, can make quite a big difference to the comfortable listening level. Extraneous sound, without your realizing it, causes your hearing faculty to turn down its sensitivity. So what might have seemed very loud, in the absence of the background noise, suddenly becomes almost inaudible.

This is another important factor to keep in mind when considering your power needs. On the opposite side of the ledger, if the unwanted sound is reverberation that comes from the sound you also

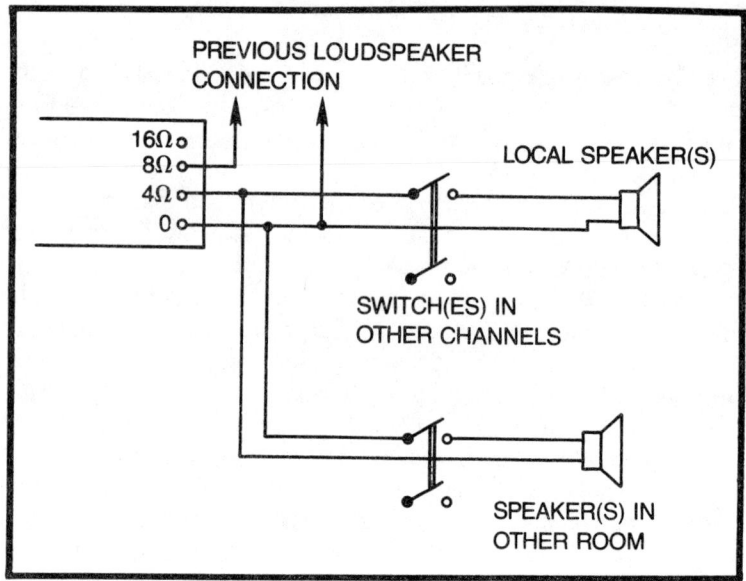

Fig. 11-7. One way of wiring extension speakers in another room.

want to listen to, turning it up will not help. It will make it worse. By turning it down, the reverberation is also decreased correspondingly, and some of the reverberation will then become inaudible to you, while you can still hear the direct sound.

There is one more factor concerned in deciding the amount of power audience absorption. We speak of audiences taking in what they are told, or what they hear, whatever it is. This is true more literally than you may realize. Human beings are very absorbent objects, and they change the acoustic properties of a room by their presence in it. Thus, if you play sound over a system in a large room with nobody (but yourself, presumably, else how else can you know how it sounds?) in it, you may need so much power.

But when a group of people begin to fill up the room, you will need more power to make the same sound audible, or comfortable to hear. This happens because of the absorption of the human bodies present as we have said. But it can also be because every person present adds some incidental background noise of his own, even if it is only breathing.

EXTENSION IN ANOTHER ROOM

A special case of the power distribution problem arises when you want to add some extra loudspeakers so you can hear in another room, and may not always want them all on at the same time.

Perhaps you would like to extend your system with speakers in the playroom from an existing system in the living room. How do you match under these circumstances?

Do you want sound in the playroom as well as, or instead of, in the living room? Or possibly both, at times? If you want both to work at the same time, sometimes, then sometimes only one or the other, the best way is to work out your matching so it is correct when both are working, because that is when you will need most power.

Switching one or the other off will not usually cause trouble then, because the remaining installation will still get as much power as it did when both were on. So how do you tackle the problem?

The system is at present connected so the loudspeakers already installed match the amplifiers. Whatever change you make must alter this so that all the loudspeakers, on together, match the amplifiers. To be able to do this, you need amplifiers with multiple impedance outputs, so you can select a tapping that will do what you want.

If your existing loudspeakers are 8-ohm units, and you install 8-ohm units in the other room, then you should connect to the 4-ohm amplifier output, and provide switches in each output (Fig. 11-7). You may find it necessary to take steps to see that the amplifier is never operated without any loudspeakers connected, because some amplifiers will damage themselves this way.

One way to do this, is to use a three-position switch, that puts one set of speakers on, both sets, or the other set (Fig. 11-8). It has no position where neither set of speakers is on. The only problem

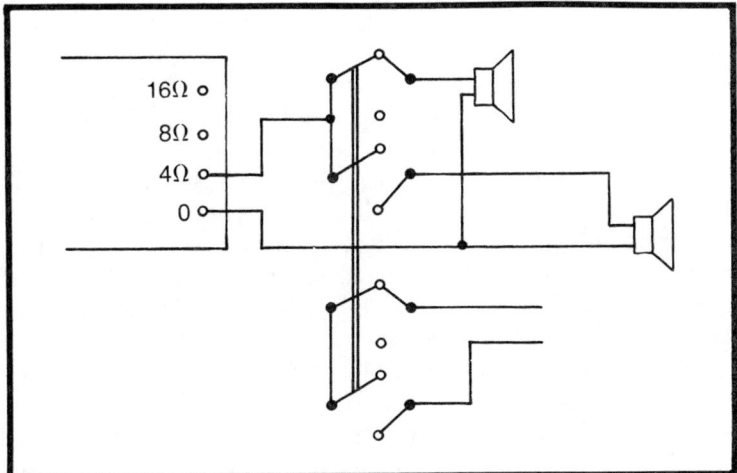

Fig. 11-8. A safer way of doing the switching.

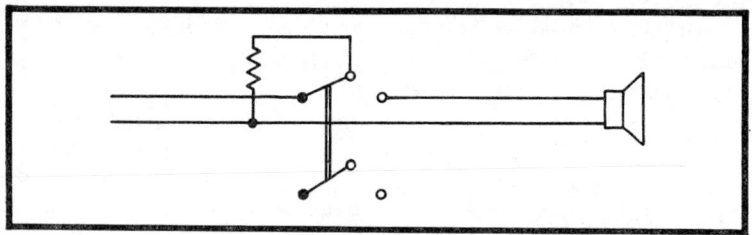

Fig. 11-9. An alternative that allows the switching to be at the remote position again.

with this is that you must have the switch by the amplifier so that, when you want to switch the remote set of speakers on or off, you must come back to the room where the amplifier is located.

One way to overcome this is to install a two-way switch in the remote room that substitutes a resistance dummy load of suitable rating when the loudspeakers in that room are switched off (Fig. 11-9). If you get 8-ohm wire-wound resistors of 10-watt or higher rating, this should serve the purpose (assuming they are 8-ohm speakers).

PHASING AGAIN

Assuming speakers connected in parallel are all correctly phased, then you connect all of the same terminals together to one side of the amplifier output, and all of the other same terminals together to the other side of the amplifier output terminals (Fig. 11-10).

If you connect them in series, you connect one terminal of the first speaker to the first amplifier terminal, and the opposite terminal of the last speaker to the other amplifier terminal, then connect opposite terminals of adjoining speakers together, step by step, between units until you have them all connected.

That assumes that the loudspeakers you use have been correctly phased during production. If they have not, or if you are not sure that they have been, you should check them to be sure. For this purpose, the easiest method is parallel connection (Fig. 11-11). You need not space the units apart, as in Fig. 10-2. For this purpose, it is better to keep them close together.

Regard one of them as the standard, and check all the others against it. If they all turn out to be correctly phased, you need not do any marking. But as soon as you encounter a difference, you should set up your own marking plan. If there is any marking on them already, choose a different marking so that you know which is yours.

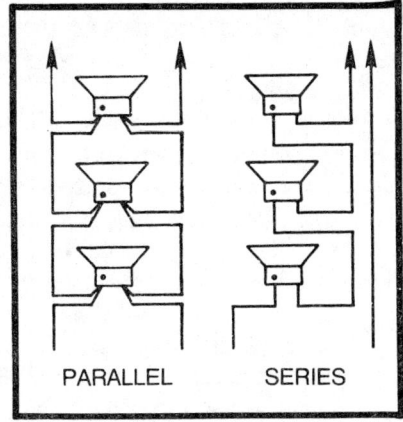

Fig. 11-10. Method of connecting in parallel or series to get correct phasing.

PARALLEL   SERIES

A common practice among loudspeaker manufacturers is to put a red dot on one terminal.

So you should use a different color paint to put on your own dots. Now, mark each unit to indicate which way around the unit adds its sound to that of the standard. Mark the terminal on the one you are testing that connects to the terminal on the standard to give the in-phase response with the same colored dot.

Telling which way is in phase is easier this way than when the units are separated, as in Fig. 10-2. When they are correctly phased, the two sounds add, giving a more bassy reproduction, than when they are not correctly phased. The reproduction sounds thinner if an

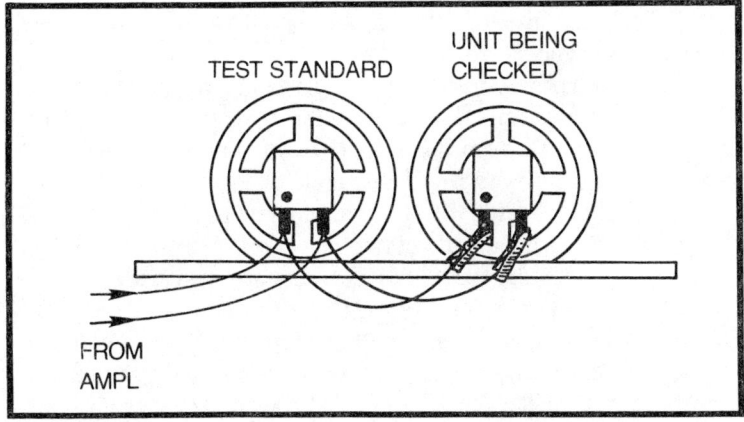

Fig. 11-11. Method of checking phasing of units fo rmultiple-unit systems, such as columns. Shown is the marking when the phasing is identical in each unit. Using alligator clips enables a whole batch of units to be checked against the one used as a standard.

out-of-phase condition exists because it lacks bass, as well as having less output altogether. And it also will seem to lose its sense of source or origin.

Having gone through all your units and marked them, you can then wire them as already described, except that when they are in parallel, you wire all the dotted terminals together, and all the undotted terminals together, separately. For series connection, you wire each dotted terminal of one unit, or parallel-connected group, to the undotted terminals of the next unit, or parallel-connected group.

When you are through, connect the line to the amplifier. In parallel, each line that goes to all of the units also goes to the appropriate amplifier output terminals for the correct impedance matching. In series, the unit terminals that get left over when you have completed your series connection go to the appropriate terminals.

MULTIWAY

So much for loudspeakers on individual channels. Now what about multiway loudspeakers as opposed to putting identical units in series, parallel, or some other combination? This involves the use of crossovers to see that the right frequencies reach the right loudspeaker units. There are some rules to learn about this too.

The first rule is that in any ordinary crossover arrangement you must use loudspeaker units, all of whose impedances are the same.

The second rule is that the combined impedance when the crossover is used is the same as that of *each* loudspeaker impedance individually. Thus, if each unit has an impedance of 8 ohms, the combined impedance is also 8 ohms.

Next, if you plan to do any changing around, you need to know the types of crossover. Each unit is fed through a low-pass or high-pass section of the crossover (or both for midrange) that allows only those frequencies to reach that unit.

Then, rule 3, the input to the various low-pass and high-pass sections must be connected, either in series or in parallel (Fig. 11-12) according to the way those sections are designed.

Having stated the rules, we will now proceed to explain them. The reason that all units should be the same impedance is because over its frequency range each loudspeaker is receiving the output from the same amplifier output tap. What the crossover really does is to provide a gradual switch from one unit to the other in the vicinity of the frequency known as crossover.

Now, the reason that the impedance is not multipled or divided, for such series or parallel connection as when you combine identical

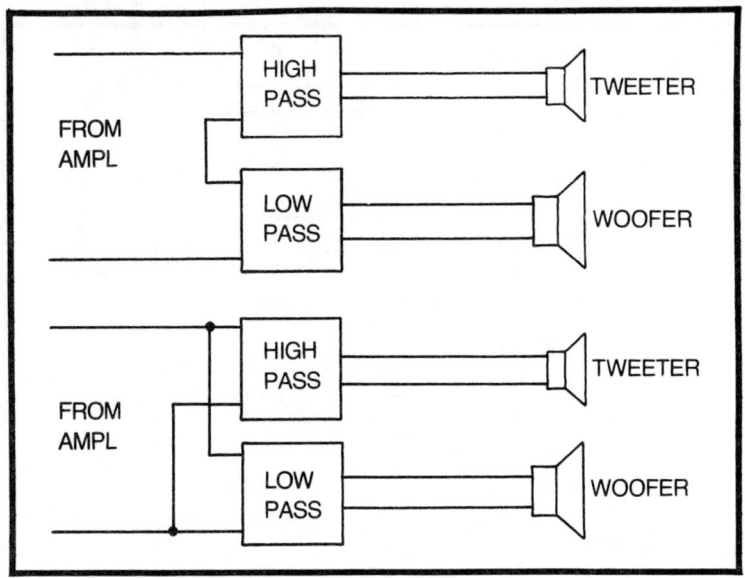

Fig. 11-12. Alternative ways of connecting different types of crossover inputs.

units without crossovers, is that at any particular frequency the amplifier output goes to only one of the multiway loudspeaker units (or at crossover frequency, it divides between them), not to both.

When you series-connect the proper crossover-section inputs, the speaker unit not receiving a given frequency has its section short-circuit that frequency over to the other section (Fig. 11-12). When you parallel-connect the proper crossover-section inputs, the speaker unit not receiving a given frequency has its section input go open circuit at that frequency, so all the power goes to the other one (Fig. 11-13). Either way, all the amplifier power goes only to the unit that is receiving it.

At crossover frequency, this is changed a little, so that some power goes to each unit, but the impedances are modified so the total still looks like one unit at the combined inputs, not two. This is the basic way to do it.

If the filters are designed for their inputs to be connected in parallel, the first reactance looking from the amplifier end is in series. The currents fed to the two filters always add up so that the combined impedance does not change if the outputs are correctly terminated, which is a theoretical condition not met with practical loudspeakers.

If the filters are designed for their inputs to be connected in series, the first reactance looking from the amplifier end is in parallel

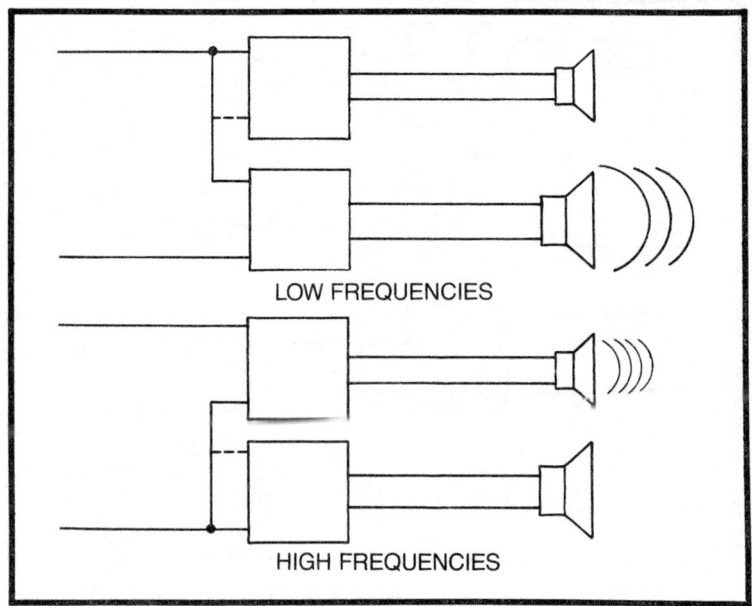

Fig. 11-13. How a crossover, designed for series connection of its inputs, separates the frequencies.

with the input. The voltages taken by the two filter inputs always add up so that the combined impedance does not change, which again is a theoretical condition that only approximates what happens in practice.

It means that, as we said in rule 1, you must have all 16-ohm units, or all 8-ohm units, or all units of some other definite impedance the same. Is there no way to get around this if you happen to have units of different impedances?

Well yes there is, but it is a little tricky. You must use a type of crossover that is designed for parallel connection at the input, and you need an amplifier with impedance outputs to suit each unit you plan to use. Suppose you have a 16-ohm woofer and an 8-ohm midrange and tweeter. You must use a 16-ohm low-pass section for your woofer, and connect its input to the 16-ohm amplifier terminals. Then you must use 8-ohm sections to feed the midrange and tweeter units, and connect their inputs to the 8-ohm amplifier terminals (Fig. 11-15).

CHOICE OF CROSSOVERS

We do not want to get into crossover design or construction in this book. But you should know a little about what makes the

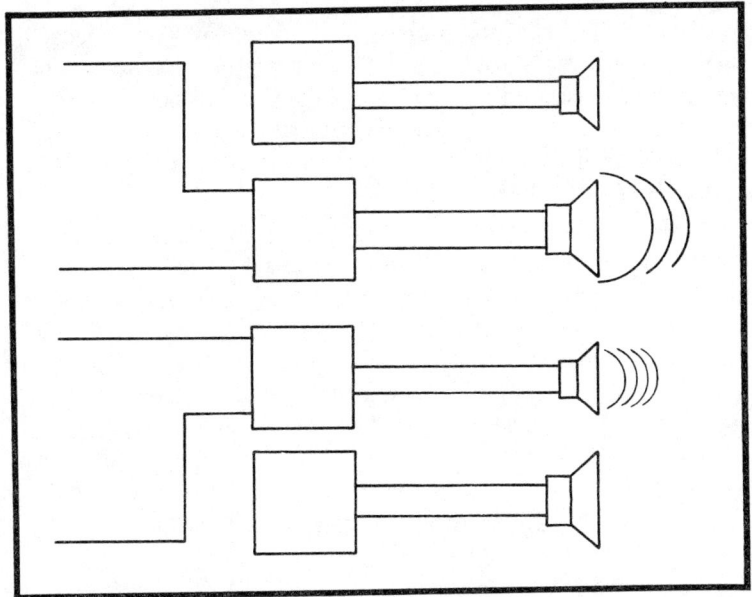

Fig. 11-14. How a crossover, designed for parallel connection of its inputs, separates the frequencies.

differences. Crossovers differ in whether you connect the inputs in series or in parallel (rule 3) as we have already said, and they can be connected only in the way for which they are designed.

But they also have different response characteristics, as well as different frequencies at which they "do their thing," called crossover frequencies. The response characteristic depends on how many reactances they have in each section (Fig. 11-16). A reactance is either a capacitor or an inductance.

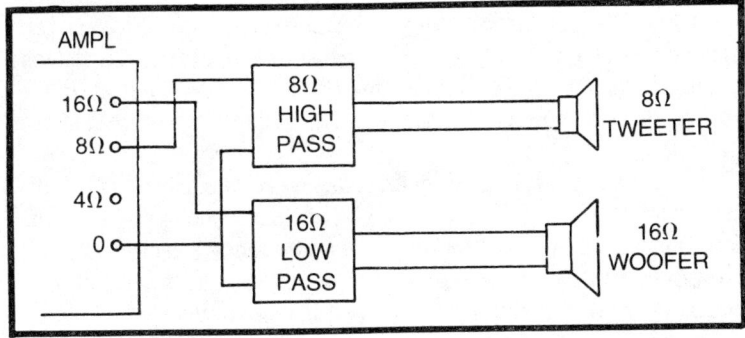

Fig. 11-15. How to use units of mixed impedances in a multiway system. It is tricky but not impossible.

The simplest kind of crossover uses just one reactance in each section, a capacitor in one section, and an inductor in the other. The next more complicated has one of each in each section. Then the next can have two of one with one of the other in each section, making a total of three capacitors and three inductors for the whole crossover. And so on.

This combines with whether the inputs are design for series or parallel combination as follows. In the single reactance crossovers for parallel inputs, the reactance is just a series inductor for the low-frequency (woofer) filters, and just a sereies capacitor for the high-frequqncy (tweeter) filters.

For series inputs, the low-frequency filter has a parallel capacitor, and the high-frequency filter has a parallel inductor.

For two-reactance filters designed for parallel input, the low-frequency filter uses a series inductor with a capacitor across the output, while the high-frequency filter uses a series capacitor with an inductor across the output.

For series-input two-reactance types, the low-frequency filter uses a parallel capacitor with an inductor in series with the output, and the high-frequency filter uses a parallel inductor with a capacitor in series with the output.

For parallel-input three-reactance types, the low-frequency filter has a series inductance at each end, input and output, with a parallel capacitor in the middle, while the high-frequency filter has a series capacitor at each end with a parallel inductor in the middle.

For series-input three-reactance types, the low-frequency filter has a parallel capacitor at each end, input and output, with a series inductor between them, while the high-frequency filter has a parallel inductor at each end with a series capacitor between them.

The same pattern goes on for increasing numbers of reactance elements. The values follow formulas that get more complicated, the more complicated the filters get. When more than two reactances are used, the values are all different. But in this book, we are more concerned with how performance varies, than with the details of their design.

The main difference between crossovers of such varying complexity is that the simple ones change over from one speaker unit to the other at a relatively slow rate as frequency changes. In any crossover, right at the crossover frequency each speaker unit gets equal power. In the simplest crossover, at an octave below or above the crossover frequency, one unit gets just four-fifths of the power, the other just one-fifth.

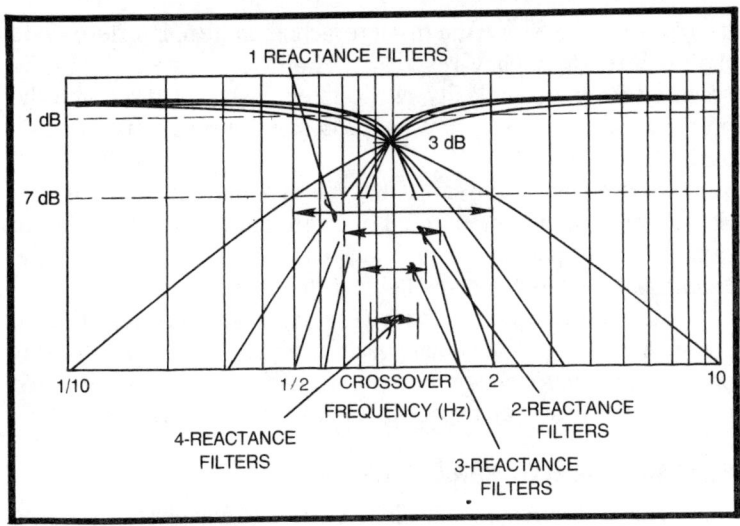

Fig. 11-16. The difference in response, according to how many reactances a crossover uses.

If you use a crossover with two-reactance sections, this brings the changeover closer together. The frequencies at which one gets four-fifths and the other gets one-fifth are only half an octave from crossover frequency.

Crossovers with three-reactance sections bring this changeover range yet closer together, making the corresponding frequencies just one-third of an octave on either side of crossover frequency. The four-reactance type (which are not too common) bring it down to one-fourth of an octave.

The uninitiated tend to think, the more the better: let's have the best crossover possible. But that is not necessarily the best in practice. The really best choice depends on many things, complicated by the fact that the impedance of loudspeaker units is not constant, but varies over a wide range, a fact that means no crossovers behave exactly as they calculated and predicted to behave.

In the action of the more complicated filters, the successive inductors and capacitors interact to produce the desired effect, about which we have been talking. But when you use a loudspeaker's impedance instead of its theoretical resistance-load counterpart, it not only has a value that varies with frequency, it is equivalent to a combination of resistance and reactance at every frequency, except a very few where it behaves like a pure resistance.

This means that the reactance part of the loudspeaker's impedance also interacts, in quite complicated ways, with the reactances designed into the filter. And this interaction, in turn, interferes with the predicted frequency response. The degree of interaction increases with the complexity of the filter, just as it does with the purely theoretical filter that uses the design value of resistance as a load.

In consequence, the simple ones behave most like their calculations, and the more complicated ones are apt to be most different from their expectations. Corrections can be made for this to some extent, but how to do this is beyond the scope of this book.

Another factor is that more reactances your crossovers use, the greater the probability that phase between the units will not stay put over the whole frequency range, even when you have connected each unit the best way by the tests described in Chapter 10.

ELECTRICAL OR ELECTRONIC CROSSOVER

Another discussion you will hear among audiophiles, concerns whether to use an ordinary (electrical) crossover, or whether to use electronic crossovers with biamplifiers or triamplifiers. This too is a little more complicated than we intended to make this book, but we can give you some answers, without going into all the theory to give you the full score!

One advantage of biamplifiers is that they reduce the possibilities for intermodulation distortion, and increase the effective total output of the system.

The reason it reduces intermodulation distortion is that it separates the frequencies that work together to produce the distortion. When a single amplifier handles all the frequencies, particularly where it is loaded with an impedance that varies like a loudspeaker's impedance does, the high-amplitude low frequencies may be slightly distorted. In a good system, the distortion is probably not noticeable because low frequencies have many harmonics in them naturally, and the distortion is just like modifying the harmonic, or overtone structure.

But the same thing causes the amplification of higher frequencies to change at different parts of the low-frequency waveform. This means that the loudness of the higher frequencies present with fluctuate at the ferquency of the lower frequency being reproduced at the same time. The effect is to make the higher frequencies sound gargled, or like someone singing under water, or with water in their throat.

This is overcome by using biamplification, because the lower frequencies go through one amplifier, and the higher frequencies through another, keeping them separate so the low frequencies do not distort the higher frequencies any more. For example, if you use three-way amplification, and each channel handles full output (which is rather an oversimplification of what happens in real life), then three 30-watt amplifiers can produce an output that would require a single 270-watt amplifier, rather than just a 90-watter.

However, that is not real power increase, just effective power. Each speaker unit will be called on to handle not more than 30 watts, using those figures.

Perhaps you can see this better, again, by figuring it in voltage. If you are using 8-ohm units, each 30-watt amplifier—make that a 32-watt amplifier, to get round figures—will produce 16 volts. Now, if one amplifier handled the whole set of frequencies, the voltages would add because the higher frequencies ride on top of the lower ones.

Thus, to get the same output as the three separate ones gave, you would need three times 16 volts, or 48 volts, still feeding into 8 ohms, whether single unit or multiway. And 48 volts across 8 ohms is equivalent to 48 squared, or 2,304, divided by 9, or 288 watts.

This is the equivalent power the amplifier could deliver if all the power was at one frequency, instead of three different frequencies combined. Actually, the amplifier will not be delivering the 270, or 288 watts (according to which set of figures you use), but only the 90 or 96 watts. But to do it from one amplifier requires a capability of delivering 270 or 288 watts, instead of the three separate amplifiers with a capability of 90 or 96 watts total.

Another advantage is that the response is made less dependent on the impedance characteristic of the loudspeaker units. If the electronic crossover is correctly designed, this is a considerable advantage. But unfortunately many electronic crossovers on the market have not been designed to take full advantage of this feature.

The way many electronic crossovers are designed is to put a succession of single reactance filters one after the other. Say it uses three such filters. Each of them is 3 dB down at crossover frequency, 1 dB down an octave one side, and 7 dB down an octave the other side. That is what we meant when we were talking about one-half power, one-fifth power (7 dB) and four-fifths power (1 dB).

Now combining three such elements in succession produces a response where the the same octave-interval points are 9 dB, with 3 dB and 21 dB on either side. Sometimes the designer adjusts this by shifting the "crossover" to the 3 dB point, instead of the 9 dB point, which means recalculating all the values.

In a true 3-element filter, the response 1 octave beyond crossover is 18 dB down. With this improperly designed one, it will only be 9 dB down. A correctly designed electronic crossover, must use feedback inside the amplifier, to get this steepening effect, just as an ordinary output crossover uses interaction between the reactance elements.

FEEDING THE FULL-BEAM MULTIPLE UNIT

Referring to the design shown in Figs. 7-1 and 7-3, that will provide a little more exercise than we have covered in this chapter on wiring the speakers to feed it. Here is our suggestion without comment. Suppose we use all 8-ohm units.

First connect units numbered 1 through 7 in parallel. That will produce a nominal impedance of 1.14 ohms. Now connect the following groups in parallel, by number: 8 through 13, 14 through 19, 20 through 25, 26 through 31, and 32 through 37. Each of those will come to 1.33 ohms. Connect all those groups in series, and the resulting impedance will be about 7.8 ohms, which is close enough to use an 8-ohm output.

Now, for power distribution, if you allow 1 watt per unit for numbers 8 through 37, that is 30 watts. The lower resistance of the inside group, will cause the seven of them to take only 5 watts, or about 0.7 watt apiece, so the total will be about 35 watts. This means the two outer rings will get about 1.4 times the power of the inner circle, which will produce a pretty good plane wave, over the 42 inches across, that will spread a little as the beam moves forward.

That is one way of doing it, but possibly not the best. If you can get it, the better way is to parallel connect, but this comes to a very low impedance. For this you need a specially built transformer. The transformer can include tappings that reduce the power to the inner loudspeakers, whether you are working with a full-beam type or a column, which will save wasting audio power in the resistor we used to effect this adjustment when we discussed this earlier.

However, that approach is not available to most home builders. A manufacturer that wanted such a system would probably have the transformer specially made; they are not generally available off the shelf. So the better way is to improvise in the ways we have been showing you in this chapter.

You may hear different stories about relative merits about different ways of doing things from what we have been telling you in this chapter. We have tried to keep it simple, and to give you the best information about facts where differences are possible. Opinions do differ, sometimes with good reason, sometimes without.

CONSTANT-VOLTAGE LINE DISTRIBUTION

Before leaving this chapter, for the benefit of those who may have occasion to use it, we will describe how the so-called constant-voltage line distribution systems work. First we need to clarify how the concept differs from conventional matching of loudspeakers to amplifier outputs.

The method we have used throughout this chapter juggles the overall impedance to match one of the amplifier output taps, and the arrangement of individual loudspeaker impedances to get the distribution right. The constant-voltage line distribution system is based on the method of distribution used for ordinary electrical power with some differences that make it necessary to explain it rather carefully.

In an ordinary house distribution system, the line voltage is 117 volts, and you connect on lamps, heaters, and other electrical appliances, which take a current depending on their power consumption. If you switch on a 40-watt lamp, it takes $40 \div 117 = 0.34$ amp from the line. If you switch on an appliance that takes 1000 watts, it takes $1000 \div 117 = 8.6$ amps from the line. And so on.

In the case of house current, there is always more current, or at least there has been until recently. Now we are told that we are nearing the maximum available. But in any event, the current we take is dependent on the total power we connect to the 117-volt line. How can we apply that to an amplifier output situation?

House current has a nominal voltage, which is held constant, to within about 6%, according to regulations. That means the nominal 117 volts may actually vary between about 110 volts and 124 volts, which is not too wide a swing. The output of an amplifier, delivering audio output that will produce sound is quite different.

But it does have a nominal maximum. For example, a 50-watt amplifier, designed to feed an 8-ohm output will produce $\sqrt{50 \times 8} = 20$ volts maximum output. That refers to a theoretical single frequency sine wave. Its peak would be about 28.3 volts, and that is about the peak voltage that the amplifier will give, whatever the average voltage and power (Fig. 11-17).

Now there is our basis for devising a different way of figuring. Although that voltage, in this case 28.3 volts, is the peak voltage it will deliver into an 8-ohm load, it is the peak voltage it will deliver into any load, using that particular output arrangement, or matching. If you changed the load to 16 ohms, it would still deliver 28.3 volts peak, or 20 volts of single frequency sine wave, into 16 ohms, which then is no longer 50 watts, but only 25 watts.

So you could take off load, resulting in a higher resistance load at the cost of losing maximum power as total output. But if that load was made up of many loudspeakers connected in parallel, like electric lamps on house wiring, disconnecting some of them would not reduce the power delivered to those left connected.

If one loudspeaker, part of that 8-ohm load, took say 2 watts out of the 50-watt total, then disconnecting some of the load would still leave that loudspeaker taking its 2 watts maximum, or 4 watts peak. This idea is the basis for the so-called constant-voltage line distribution system. We should point out that calling it constant voltage applies not to the voltage measured instant by instant, but to the fact that, at any particular instant, the voltage will be the same at that moment, regardless of how many or how few loudspeakers you have connected.

Thus at one instant you may have 20 volts, at another instant you may have 2 volts. But the voltage at those instants will be the same, regardless of how many loudspeakers you have connected.

Now to work on this scheme. It is not practical to match the whole amplifier to an impedance such as 8 ohms because that would involve the use of rather heavy wiring to carry the heavy currents involved. So, like the house current situation, we pick higher voltages, and make the impedances fit the voltages we decide upon.

Two standard nominal voltages are used for constant voltage line distribution systems: 70 volts and 25 volts. Actually, the 70-volt figure would be more accurately stated as 70.7 volts. It is chosen because 70 squared, which you need to calculate the power into various impedances, comes to 5000—at least 70.7 does.

It is also chosen because the peak value of a sine wave that reads 70.7 volts is 100 volts. So this means that, on the 70-volt output line, an amplifier delivers an output that reaches 100 volts peak.

One more limitation. We said that the amplifier gives that output, whatever impedance is connected, provided it is higher in value than the rated one. Where it was 8 ohms, it would deliver a 20-volt sine wave, or 28.3 volts peak into any impedance larger or higher than 8 ohms. If you connected, say a 4-ohm load, then the amplifier would distort badly, before it reached the 28.3 volt peak.

This is true for the constant-voltage line distribution system too. You must be sure to keep the total load within the power rating of the amplifier, or above the impedance rating that corresponds with that power rating. Let us see what that means.

Suppose the amplifier is rated at 200 watts and has a 70-volt line output. This means the impedance for delivering maximum power

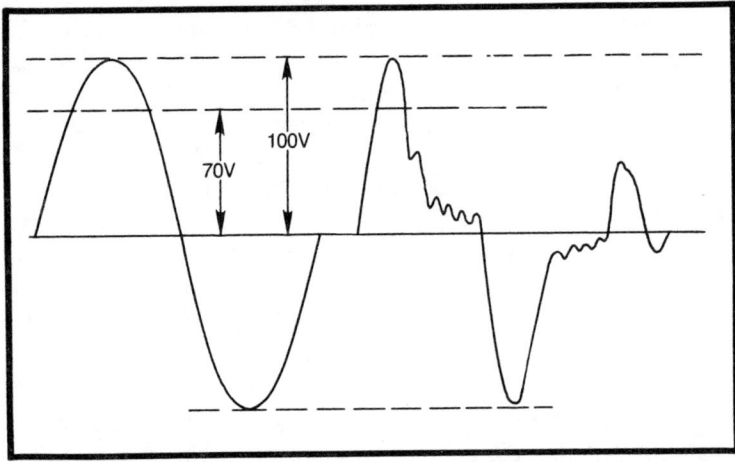

Fig. 11-17. Waveforms to show meaning of so-called 70-volt line. At left the theoretical sine wave output on which the 70-volt designation is based. At right a sample program waveform, showing that the correspondency is at the peak, or 100-volt point.

must be $5000 \div 200 = 25$ ohms ($R = E^2/P$). Into any load impedance higher than 25 ohms, this amplifier will deliver its 70-volt sine wave, or 100 volts peak of any waveform.

Now, how do you make up such a load with conventional 8-ohm, 16-ohm, or whatever they may be, loudspeakers? For this you need constant-voltage line matching transformers. These transformers have a primary and a secondary. If they come with the loudspeaker as part of a unit equipped for connecting to a constant-voltage line, they may have a secondary designed just for the impedance of the loudspeaker on which they are mounted. But if they are universal constant-voltage line matching transformers, they will have a secondary with tappings designed for different loudspeaker impedances, such as 4, 8 and 16 ohms (Fig. 11-18).

The primary, on the other hand, has tappings that are not marked in impedance, but in power ratings. Thus a transformer primary may have a common terminal, and tappings marked 5 watts, 2 watts, 1 watt, and 0.5 watt. What does this mean? Actually the transformer is designed on the basis of impedance and merely labeled on the basis of power rating.

If the amplifier delivers a 70-volt sine wave, then to give 5 watts, the loudspeaker must present a load on the primary side of the transformer of $5000 \div 5 = 1000$ ohms. To give 2 watts, the primary impedance must be $5000 \div 2 = 2500$ ohms. To give 1 watt, the primary impedance must be $5000 \div 1 = 5000$ ohms. And to give 0.5 watt, the primary impedance must be 5000 ÷ 0.5 = 10,000 ohms.

Those are nominal watts, based on that theoretical 70-volt sine wave. Whatever *actual* power is delivered can reach double that marked (or computed) value because the *peak* voltage is 100 volts.

Now, your 200-watt installation can be made up of any number of loudspeakers connected through their individual constant-voltage line matching transformers. The total rated power of all the speakers must not add up to more than 200 watts. When it adds up to just 200 watts, then the amplifier load will be the minimum allowable 25 ohms in our example. If it adds up to less than 200 watts, the amplifier load will be higher than 25 ohms.

Each will get its proportionate amount of the 200-watt sine wave, if that was fed into the system, or of the 400-watt peak program material, which would consist of a complex waveform. The thing to watch is that you do not overload the amplifier by connecting more than 200 watts' worth of speakers which would load the amplifier with an output impedance of less than 25 ohms.

Now, if you examine an amplifier designed for this kind of service, you will find that the 70-volt outputs and the 25-volt outputs each share a common center tap, which is grounded. This is as a precaution against system instability. If one side of a 70-volt system was grounded, the other side would reach 100 volts peak at maximum output. In audio, that is a lot of volts to be "floating around" a building.

By using a center tap to ground, that 100 volts peak consists of 50 volts peak each way to ground. When one wire is momentarily 50 volts positive of ground, the other is momentarily 50 volts negative of ground, and so on. This means that the high-voltage field is neutralized, because the two wires are close together in the same cable wrap.

Just as the 70-volt output represents full power in a single frequency sine wave, so does the 25-volt output if it is fully loaded. And as the 70-volt output is equivalent to a 100-volt peak output, so the 25-volt output is equivalent to a 35-volt peak output. But the ratings are calculated on the basis of 70 volts or 25 volts.

A transformer designed to feed a loudspeaker a rated 5 watts presents a primary to the constant-voltage line that looks like 1000 ohms as we calculated earlier. Now, if that same transformer with its loudspeaker is connected to the 25-volt line, it will accept a rated 25 squared, divided by 1000, which is 0.625 watt, just one-eighth of the power it will take from the 70-volt line. Now, there are several ways

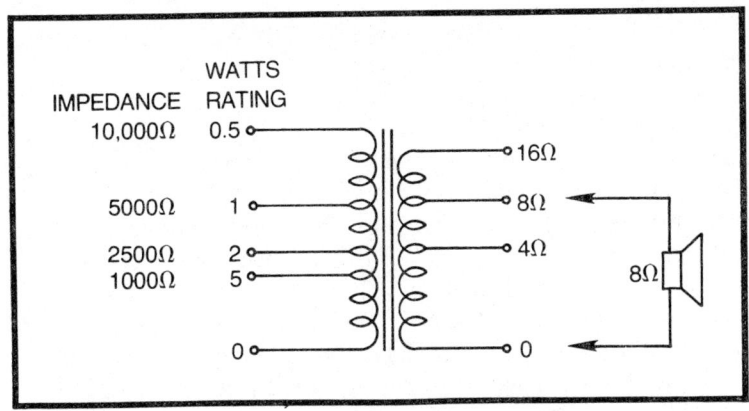

Fig. 11-18. A typical universal constant-voltage line matching transformer.

in which the constant-voltage line can be used. That is the purpose of it: to give flexibility in delivering power effectively.

The 70-volt line is, in general, for high-level distribution systems, where each loudspeaker is to serve a fairly large section of audience. The 25-volt line is for low-level distribution systems, where a great many more loudspeakers are used, so that each one serves only a few people, all of whom are relatively close to it.

Whichever general method you adopt, some units will serve larger areas than others, because of the configuration of the place. For example, units feeding into corners may want more, or less, power than those feeding straight side sections. You may figure this by looking at the layout, or by trial and error, after you have the speakers in position.

But you need to figure on using taps so that the total of the rated watts is equal to the rated output of the amplifier if you use the 70-volt line for all of them. If the two numbers do not match, the total of the loudspeaker ratings should be less than the amplifier rated output, not more.

Now, how do you use the 25-volt line? You can figure this in a variety of ways. What using the 25-volt line does is to cause the units connected to it to take one-eighth of the rated power, according to the way they are connected. So you could add up the rated power of the units connected to the 25-volt line, and divide by eight, to find the power they will take from the amplifier.

Thus, if you connect the whole system to the 25-volt line, your 200 watt amplifier will serve loudspeakers rated to a total of 1600 watts. This does not mean they will get 1600 watts, but that they will share the 200 watts the amplifier actually gives equally.

Alternatively, you may connect part of your system to the 70-volt line and part of it to the 25-volt line. This will mean that you add up the wattage ratings of the speakers connected to the 70-volt line directly. And you add up the wattage ratings of the speakers connected to the 25-volt line, and divide them by 8, before adding them to the total *demand* on the amplifier.

Then you need to be sure that the total demand is not more than the rated power of the amplifier. If it is, you will get distortion. Actually, it may sound louder so connected. But it will be distorted, and thus less intelligible. And if you have other sounds to overcome, although it sounds louder, it will not be really, so you will lose.

One more thing you may need to attend to is the possibility of needing bass-cut capacitors when you use horns mixed with cabinet speakers on the same system. Cabinet speakers can be fed the full frequency range, but horn speakers must be protected against being fed frequencies below their cutoff frequency, which may be 200 hertz or 400 hertz.

The following table gives values that can be used for this purpose, as well as the wattage ratings for the 25-volt line, as well as the 70-volt line.

70-Volt Line	Primary Impedance in Ohms	Series Capacitor (μF) for		25-Volt Line Wattage
		200 Hz	400 Hz	
0.5	10,000	0.082	0.039	0.063
1	5000	0.15	0.082	0.125
2	2500	0.33	0.15	0.25
5	1000	0.82	0.39	0.625
10	500	1.5	0.82	1.25

Smaller values of capacitor can be used, if desired, but larger ones are not desirable.

We should make some comments on adjusting such a system. Earlier in the book, we pointed out that 3 dB is a barely discernable difference in level, which is true, when you are comparing levels, one against another, played separately. But when you are attempting to fill an area with sound, it is better to think in terms of relative power.

All of the loudspeakers that adjoin the area they serve share in the job of filling that area with sound. So if one of them works at a 3 dB high level than the others, which is twice the power, it will do more than its share toward filling the area with sound.

This means that, in setting up a system like this, a 2:1 error in relative power levels can make quite a difference, although a 3 dB change in level may be barely noticeable, under other circumstances. Under these circumstances, it becomes a question of where the sound seems to come from, at any particular listening location.

Because of this fact, as well as careful figuring out before hand, you may find yourself having to make adjustments after installation to get the best possible distribution. This means you may find it desirable to work on tap changing while the system is live—actually delivering sound to the audience area.

Of course, you will probably be doing this with the audience area empty, which is one thing for which you will want to make allowances: how it will change when that same audience area is full of human bodies with their attendant acoustic absorption. To judge this, you need to note how absorptive or reflective it is without them. The more reflective it is without them, the bigger difference they will make by being there.

If we don't warn you, you will soon find out that 70-volt line can give you quite a shock when it is delivering audio power. But while 70 volts of audio can give you as sharp a kick as 117 volts of a 60 hertz supply, it is far less lethal. Audio frequencies "throw you off," while the steady 60-hertz supply, if high enough in voltage, can grab your muscle reactions, making it difficult to let go, which is how a number of people get killed by it.

For all that, if you happen to be working at the top of a ladder, changing connections to a loudspeaker located in the roof, you do not exactly want to be thrown off either. And there is another factor. With a 60-hertz supply, unless you get into real high voltage, you must touch both sides to get the shock. Of course, you can get a shock from the live side to ground, but you must *contact* both live and ground to get the shock.

With audio, because of the higher frequencies involved, you may not need a ground in the same sense. If you touch one high-voltage side, your capacitance to ground may be enough to give you a shock without a direct path to ground. This means that, to avoid getting a shock from audio, you need to stay away from the cable conductor that carries the audio. Handle it by the insulation only when it is live.

You can make checks by touching the live wire on alternative terminals, while someone listening reports to you on the relative changes in sound distribution. Then, when you think you know how you want it to be, have the sound turned off while you make the

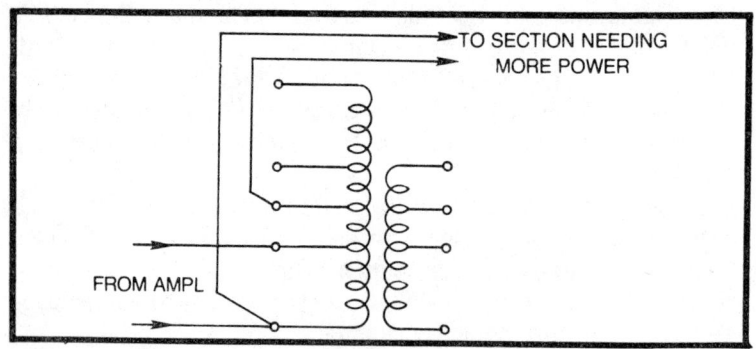

Fig. 11-19. How to use the primary of a single transformer to try out changes in level to whole sections of loudspeakers before making the permanent change individually.

connection. You do not have to turn the amplifier off, just the sound. That zero terminal in the middle is ground, and when there is no sound, both the live wires also drop to ground.

Although the constant voltage line system seems to offer a maximum in flexibility, there may be occasions when you "run off the end" one way or another: it seems as if you need one more tap in the direction you are going.

Sometimes when this happens you can "steal" a bit more by using a change in secondary taps. If your speaker unit is of 8-ohm rated impedance, and the transformer secondary provides for 4, 8 and 16 ohms, you would normally connect it to the 8 ohms tap. But if you find that the highest wattage rating on the primary taps is not quite high enough, you could connect the loudspeaker to the 16-ohm secondary tap.

And if you find that the lowest wattage rating on the primary taps is a little too loud, relative to other speakers in the system, you could connect the loudspeaker to the 4-ohm secondary tap.

One more thing: if your problem is one of changing a whole lot of speaker taps, relative to a whole lot of other speaker taps, you can shorten the work a little sometimes by using the primary of a line transformer to effectively change them all at once (Fig. 11-19). Then, having found what is needed, you can go along the line and change them all to correspond.

If you go in on, say the 5-watt tap, and come out on the 2-watt tap, you are actually stepping up the voltage. This means that, to get the same effect on each transformer, you need to go down, so the units take more current at the same voltage. This needs some careful working out, but if you work out what you are doing, it can save you some time.

12

Making Good Connections

Now, you have got all those technical details straight, you know what to get and how to put it together to make the kind of installation you need. All you have to do is do it. Or perhaps you have done it and are having problems. That is what we want to pick up in this chapter.

If you choose a system that comes as a number of electronic units with a foolproof set of connecting leads, all provided with the necessary plugs and sockets, there is very little that can go wrong. And if something does, there is the warranty. But if that is all you wanted, you probably would not have bought this book in the first place. You wanted to do something original that you figured out for yourself.

So it comes down to the question of how you make reliable connections and, if something does not work right, how you track down what is wrong. You may be wiring up your own system completely, or you may be adding to an existing system. Either way, you want to do it right.

CONNECTING WIRE

Most of the system installations we have covered in this book require the flexibility of amplifiers with different impedance output terminals, and the ability to change the phase of connections to loudspeakers. Where this is not necessary, of course, you can buy the ready-made audio leads, with the plug molded on to make perfectly good connections.

But we are assuming you need some connecting wire that you will have to make reliable connections with, and are going to have to

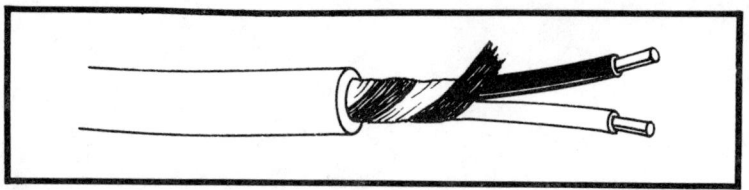

Fig. 12-1. Color-coded wire helps in keeping track of phasing during wiring up of a system.

do it all yourself. First, you need a suitable grade of wire. Loudspeaker connections need to have low resistance, or you will lose some of your power in the wiring, instead of delivering it to your loudspeakers. So it should be of a good heavy gauge, preferable stranded. It should also be well protected with insulation.

Loudspeaker wiring carries electrical currents of a sort—audio currents are electrical—but they are not dangerous to humans or animals as are house currents. However, any damage to the circuit may not be good for your amplifier. Many amplifiers can be permanently damaged, even blow their output transistors by either a short circuit or open circuit. Even if they have automatic protection circuits, as some of the best ones do, that kind of treatment does not do them any good.

So your wiring needs protecting from the ravages of humans and animals around the house, rather than vice versa. Watch for possible causes of such trouble when you are routing your wiring, about which we will have more to say a little later.

Get some wire with multistranded conductors, equivalent of 16 gauge or 14 gauge at least, to keep the resistance low. It will save a lot of phase checking if you get wire that has the insulation, or the wires themselves color coded in some way (Fig. 12-1). Much wire of this kind achieves this by using different colored insulation inside the outer sheath. Another form, sometimes available, is the flat strip, that comes with transparent vinyl insulation. One set of stranded wire is tinned, while the other is bare copper, making a difference in color.

It is true that connecting a loudspeaker either way will make it work, but the importance of connecting with consistent phasing was described in Chapters 10 and 11.

MAKING CONNECTIONS

In your anxiety to get the system working, and to prove it out, you may twist connections together, or hook them onto terminals, just to get the music through. But because loudspeaker impedances are low, you need to make sure your connections are very good, otherwise the performance may not be right, and your connections could later get noisy. This means that soldered connections are best.

However, if your amplifier has screw-type terminals for the loudspeaker outputs, watch carefully to be sure that you make a good connection. This means that each connection should be secure, and that there should be no possibility of their shorting across. When you strip the wire of its insulation at the end, it is best to use one or other kind of the proper wire strippers.

Strip enough to make a good connection, but not enough to allow the wire to short across to other terminals. Take care that you strip off just the insulation, without cutting any of the fine strands of wire. Then watch for stray strands of your multistranded wire. We specify multistranded for two reasons. Solid single-stranded wire is not heavy enough to carry the loudspeaker currents without losees, and it will not stand up to being flexed or moved like multistranded will.

A way to prevent stray strands from getting separated from the bunch, is to twist them more tightly (Fig. 12-2) at the end. Watch this all the way, because a stray strand can touch another terminal, causing a short that can mean trouble in more ways than one. The obvious one is that, if you short out your loudspeaker line, you will not get any sound on that loudspeaker. But the short may also damage your amplifier, or at least put an unnecessary strain on it.

So be very careful in making such connections to see that they are neat, secure, and clear from one another.

SOLDERED JOINTS

When you make a soldered connection, you need to take the same precautions: to make a good connection, and to avoid possible short circuits. To make a good connection, you need to be sure that the terminal to which you will solder, and the wire end that you are going to solder to it, are well tinned with solder. Only solder them together after you have made sure that solder has run freely over each one separately (Fig. 12-3).

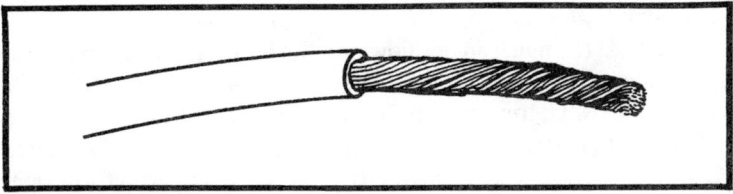

Fig. 12-2. Tight twisting of the multiple strands helps stop them "straying."

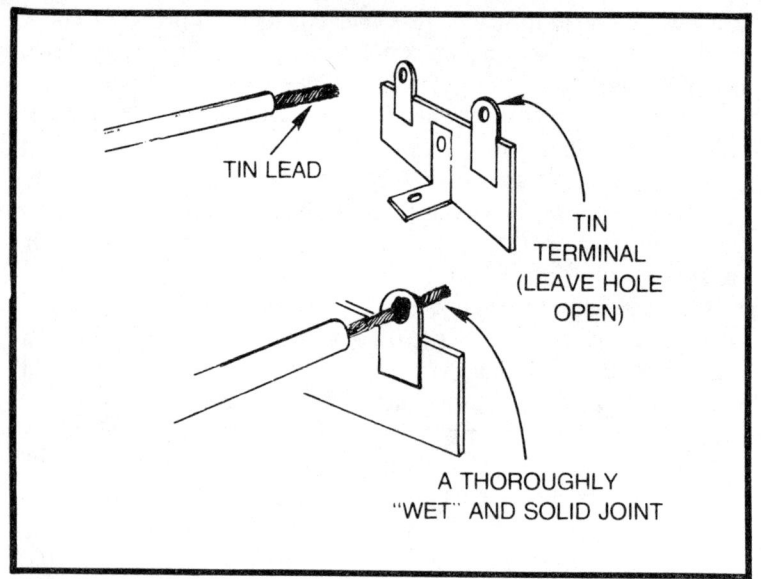

Fig. 12-3. How to make a good soldered joint: tin both sides before bringing them together.

To remedy this, take a little fine sandpaper and clean the metal of the terminals, or the individual strands of the connecting wire, until they are bright. Then apply the resin-cored solder to tin it properly. For all soldering in electrical or audio circuits use resin-cored solder, either 50-50 or 60-40 grade.

Twist the strands of the connecting wire together, just as we said for using screw terminals, before you finally tin them when making a soldered connection. Having tinned both sides, you can now make the connection and finally solder it together so the joint will not oxidize to cause a bad connection later.

You need to learn the technique of making a good soldered joint, and be aware of what can cause it to be a bad one. We have already mentioned the tinning of both pieces before you bring them together. Then you put them together, and make the joint in the same way, so the solder flows freely just enough, not too much, to make a good joint.

Letting too much solder flow into the joint can make a blob that may endanger shorting over to an adjacent piece of metal or other connection. Not letting enough flow in may leave the joint inadequate so that it will quickly deteriorate into a poor joint. Then you need to learn to make the joint in a way that all the parts are at the right temperature at the right time.

In each of these steps, you need to bring three things together, the part, the solder gun or iron, and the solder so that heat reaches the melting point of the solder just right (Fig. 12-4). If you heat the joint too much before you apply the solder, the joint can get too hot to make a good connection. And if you heat the solder, so it reaches the joint before the part is hot enough to receive the solder, the resin core may evaporate off or deteriorate before it can do its job.

Depending on the size of the solder gun or iron and the size of the part or wire to be soldered, you need to adjust your timing a little bit for best results, but the principle does not change. First be sure your iron is in good shape for the job. If it is at all corroded, clean it off with a file, down to bright copper, and tin the iron first.

With constant use, an iron corrodes. The fluxes used to prevent oxidation of the joints you make, acting continuously on the hot iron, erode their way into the copper, so that the solder forms a small pool, usually on one side of the point. Then this pool of hot solder gets too hot, causing it to disintegrate into its component alloy metals. Now, when you add more solder, it does not flow properly, and you have difficulty making a good joint.

The way to protect yourself from having this happen, is to keep the point of the iron or gun smooth and pointed at all times. As soon as you notice that the solder is beginning to form a recess in the metal of the point, file the metal away, and tin the iron so you have a good working point.

When you tin the iron, have the resin-cored solder ready to apply as soon as you have finished filing it down to shape. The bare

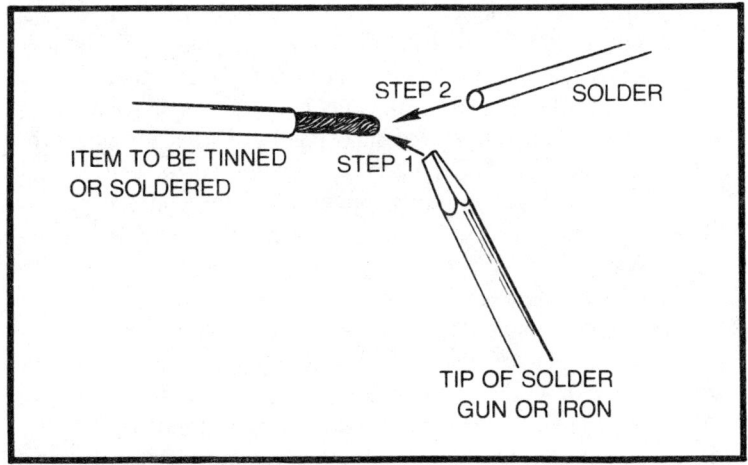

Fig. 12-4. Sequence to observe in making a good soldered joint.

copper will oxidize very quickly, going from the bright yellow to a dark red in a few seconds if it is hot enough to tin. So you must get the solder onto the bare filed copper before the metal darkens in this way. Once it is on there, you will have a well-tinned iron to make good joints for a little while.

Then watch for deterioration, and do it again, as soon as you notice the beginnings of a hollow forming.

Now, for each operation of tinning, or making a joint, you need to bring the solder in contact with the part, just as it is reaching the melting point of the solder, so that the resin-cored flux runs into or over the joint, quickly followed by the solder itself. This ensures that you have a good flow of solder right where it is wanted.

This means you apply the solder gun, or iron, first to the part to be soldered, long enough to heat it about right, then apply the end of the resin-cored solder to both, until enough flows onto the joint to make it good, but not too much.

Another thing to watch for, in making soldered joints, is crystalization of the solder, which makes another kind of bad joint. If the solder is kept at its melting point for too long, and especially if it is made too hot, it breaks down into its separate alloys, lead and tin, and will no longer set in the same form as it was before it melted.

The solder should get just hot enough to flow freely, just long enough to flow where you want it to go, and no more. Then the resin core will flow ahead of the solder to keep the parts from oxidizing before the solder gets there. Such oxidation, which is caused by the heat, will prevent the solder from properly "wetting" the parts, making another kind of bad joint.

When you are soldering the end of flexible wires or cables, another thing to watch is that you do not leave a point along the wire where the fixing of the wire creates a stress on the next piece of wire that will eventually cause a break (Fig. 12-5). To avoid such breakages, it is good to provide some kind of mechanical anchorage for the cable so it cannot bend at the point where the solder leaves off.

As well as solder spreading over the joint you are making beyond where you need and want it, which we have already mentioned, you need to watch for stray solder getting somewhere else that it should not be. It is very easy for a little sliver of solder to drop down and make a short circuit in another part of the circuit.

Usually, such a droplet can fairly easily be removed without the use of the iron or solder gun because it sets as soon as it touches a cold part. But it still makes contact, even though it is not a good joint. You can usually remove it, either by pulling it away or by peeling it off.

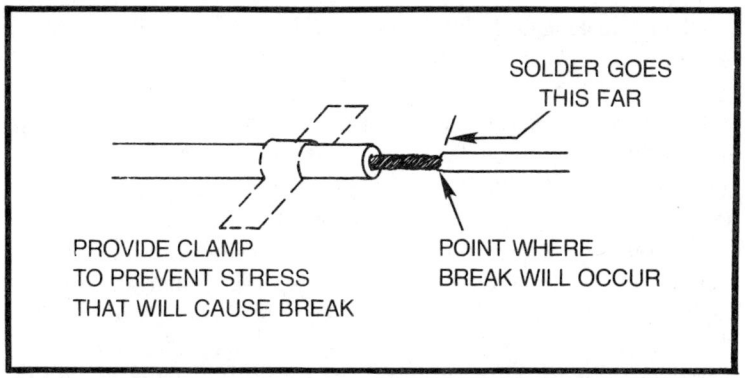

Fig. 12-5. One way in which solder can increase danger of breakage, and how to avoid it.

With joints you make that seem good, it is good to test them to see whether they will either pull apart or peel apart. One good practice is to make a joint that is mechanically good by wrapping the wire around the terminal to which you are going to solder it. But also take care to see that it is electrically good by being sure the solder flows into the joint properly. One way to check this is to try to unwrap the joint, after you have soldered it. If it will not unwrap, you have a good joint.

EXTRA CONNECTIONS

A special problem you may encounter is where you want to break in on an existing system that has neat plug-in connections; there is no ready access for additional connections. One way might be to use an appropriate combination of multiway adapters. Another would be to bring out additional sockets on the amplifier by getting inside, drilling holes and mounting the additional sockets.

This is something that will have to be figured out when you look to see what is possible. If the amplifier is very tight on spare space, the better way may be to buy one of those small aluminum boxes, and mount the necessary interconnection sockets on that, using plug-in leads to go from the amplifier to the box, from the box to the existing loudspeakers, and from the box to the newly installed loudspeakers. The box can also house a switch if you need one.

Such a box can be used to house any of the switching arrangements shown schematically in Figs. 11-7, 11-8, and 11-9. What needs care is the planning of the interconnecting leads. Of course, you could wire them in permanently, making soldered connections. But often the nicer way is to make them plug-in. Then you have a problem of making them foolproof, or at least so you will not forget how they were meant to go.

You can buy audio leads at your local electrical or audio shop, along with the necessary sockets to use with them. Remember in selecting your purchase that you need to make a low-resistance connection, because loudspeakers have a low resistance and if resistance develops you will have a noisy connection.

If you are using stereo or quadraphonic, you need connections for every channel, which adds up to quite a lot of ins and outs. Figure out a neat layout that will make the goupings of sockets reasonably obvious. Then label them all carefully, such as *from amplifier, to living room speakers*, and on, as well as *left front, right front*, and so on.

WIRING

Another thing you need to take care with is routing the connections so they will be adequately protected. When you first set up a system, you probably trail the wires across the floor, through doorways, up or down stairs, and any way that gets them where you want them to go. Now you have the problem of making a permanent installation: putting the wires where they will be safe.

You can lay them under carpets sometimes. This is where the flat twin type of cable is convenient. Or you can place them behind beadings and through walls or paneling to get them out of sight and protected. Always look for the best "way to get there" that will involve the least trouble and the best safety for your connections.

In a complete system, you will have a lot of connections to make, not just loudspeaker connections. Each has its own problems of getting where you want it to go.

If you are connecting a permanent system, you can take time and make permanent connections. But if you are putting together a portable system, you need plug-in leads for everything. Then you need to make sure that your system of plug-ins is foolproof.

Do not make the mistake in selecting your connectors, with their different plugs and sockets, of thinking that you will remember how it connects together. Once it has been in use, or even not in use, for some time, you will probably forget how it was supposed to go. When that happens, you want to be able to just put plugs into mating sockets and be sure it will work without having to figure out from scratch how it was supposed to go.

So work out a system, in order that every kind of connection uses different plugs and sockets, making it impossible to make a wrong connection of any kind.

Particularly, you want to make it impossible to connect a loudspeaker to a power socket, or an amplifier input to its own output. There are only a limited number of variations in connectors available, so you may have to settle for some possible wrong connections. But figure out what would be the consequences of such a wrong connection, supposing you or someone else should make it accidentally.

For example, if you were to connect a loudspeaker to a power socket, you would burn out its voice coil? Or if you get the amplifier's output connected back to its input, for example through an auxiliary input intended for radio or tape, it will oscillate violently?

ROUTING LEADS

In making connections, you also have to figure out where to run them. As we have said, to get it working you probably just trail them across the floor. But now, having made sure your idea works, you want to tidy it up. If you use flat twin, it is easy to lay under the carpet. If you don't have carpet in the room where you want to install, you will have to think of something else.

If your room has beadings that run round the doors, along the base or top of the walls, etc., maybe you can use these in some way. Some beadings or trims have space underneath—they have a hollow underside—in which case you may be able to utilize that space to run your wiring in, out of sight, and out of danger.

You also have to think about how to get the lead into, and out of, such a running space, where it continues to its respective destinations. You may need to make small notches on the underside, or edges of the beading or trim so then when it is secured in place, it does not cut into the insulation of the connecting leads.

Another place where extra attention is needed is where you want to go from one room to another, through a door, or perhaps through windows. If the door is only a doorway, with no closeable door, your problem is no more complicated than finding a way for the lead to cross any other space (or go around it). But where the door is a closeable one, you need to think what happens to the lead when the door is closed.

There may be an adequate clearance underneath under the carpet that also goes under the door. But if the door has a sill, and if it is designed to be draftproof or even weatherproof, you need to do a little more thinking. Maybe you will need to unmount the sill or whatever, make a track for the lead, replace it all, and then caulk the channel where the lead enters and leaves.

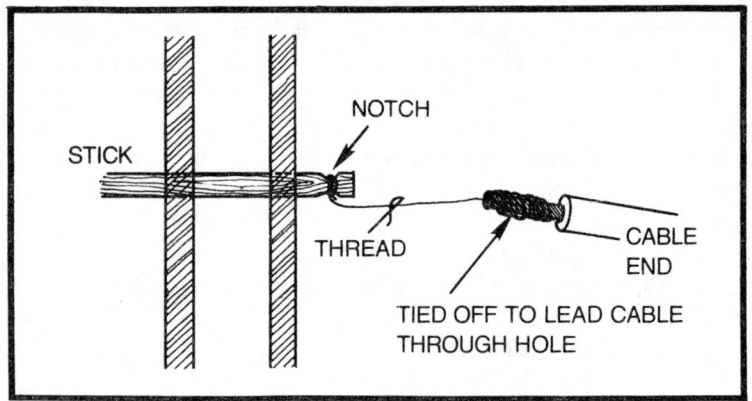

Fig. 12-6. How to get a lead through a wall, when you have to do so.

Sometimes the most economic way to get from one room to another may be through the wall, rather than round it. If your equipment is just the other side of a wall from where you want the loudspeakers it is to feed, and the path around would be a lot longer, you should find a suitable spot at which to drill the wall, and put the lead through.

In doing this, there are several things to think about. First, how you will get the lead or cable through? Then how you will remove the unsightliness of the hole, at some future time when you no longer want the cable going through there?

Walls are seldom solid. They have vertical studs, at which point they are virtually solid, but between studs they are hollow. The space between may or may not contain insualtion. Inner walls usually do not. If you drill where there is a stud, you have a long hole to drill, but threading the lead through will be easy when you are through.

If you drill holes in each side where there is not a stud, you have to get the lead through both of them. If you poke it from one side, it will get lost in the space between, and you will find it almost impossible to get it through the second side of the wall.

To achieve this, you need something stiff to pass through the wall from the opposite side first (Fig. 12-6). Then attach something slender but flexible, such as a thread, to the stiff stick, and tie that to the end of the lead, taking care to bind it so it leads the end of the lead into the hole. Now withdraw the stick from the hole, pulling the thread through, and use the thread as a pilot to pull the lead through.

Don't forget the last part of selecting your location for going through the wall: what you will do when you no longer want the lead through there. Select a spot that is not too prominent, on either side, and where the holes can easily be filled so the fact that there ever was a hole can be completely covered.

FAULT TRACING

Now, suppose you have done all this, and your system does not work for some reason. Or it has been working and quit. How do you go about finding what is wrong and correcting it? The most likely thing is a bad or missing connection. So check your connections thoroughly. First do so visually. But then, if it all looks good, do not take anything for granted: the fault has to be somewhere.

A good flexible lead will not usually fracture inside the insulation, but it can happen, especially if the lead has been given a lot of repeated bending. This will usually happen near to one of the ends, and a good way to feel for it, without going through the full fault-tracing procedure, is to pull the leads carefully.

If you find one of the leads seems to be elastic, it is because the wire inside is broken and the insulation around it stretches.

If you cannot find the fault in one of these simple ways, you will have to resort to some fault-tracing procedure. For this you can use, either a volt-ohm-milliammeter in its ohms position, or if you do not have such a meter, you can improvise with a flashlight. How to do this will depend to some extent on the design of the flashlight. It is easier if the case is plastic.

What you have to do is to bring out a couple of wires that do the same thing as the pushbutton that makes the light come on (Fig. 12-7). Then you trace the circuit to find out where the break is.

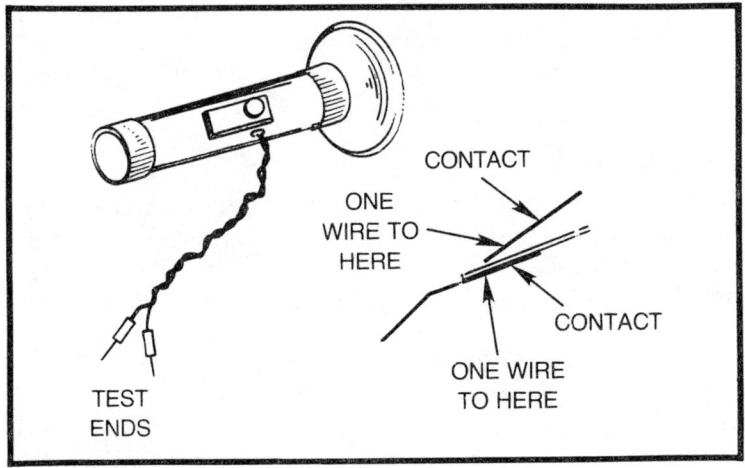

Fig. 12-7. Using a flashlight to make an improvised continuity tester.

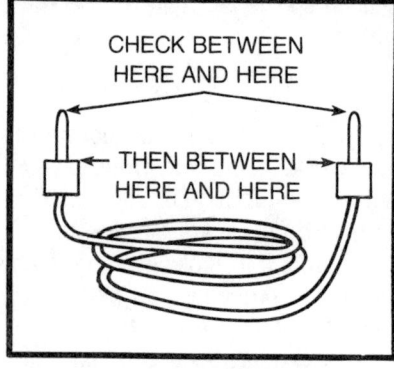

Fig. 12-8. Testing continuity of a plug-in lead.

Where you have plug-in leads, you can remove the whole lead, and try for continuity on each side to find which one is not making the connection (Fig. 12-8).

With a permanent installation, if the loudspeaker resistance is high enough to make a difference in how the light comes on (that is, it will be dimmer when the loudspeaker voice coil is in series with the lamp) then you can check the circuit without making any disconnections. Also, doing it this way, when you check through the voice coil, you will hear a plop in the loudspeaker that will give you a second check that you are getting through.

If neither the flashlight comes on, nor the voice coil makes a plop when you connect directly across the loudspeaker terminals, then your problem is in the loudspeaker itself, not in the wiring from the amplifier.

And if the circuit is okay, from the amplifier out through the loudspeakers and back, but you still get no sound from the amplifier in the loudspeakers, something must be wrong with the amplifier, or the connections at the amplifier end.

CABLES TO USE

We have commented on ways to assure phasing, and how using an appropriate cable can assist in this. But there is another respect in which choice of cable is important. For installations wired at voice coil impedance, the important thing is to use a heavy enough conductor.

Preferably it should be flexible, using multistranded wire. But it should have an equivalent cross section of at least 16 gauge wire, preferably 14 gauge.

When you are making a constant-voltage line installation, such heavy gauge wire is not necessary; insulation is the important thing.

For conductor size, bell wire would be adequate for this purpose, but the insulation would not be.

Again, you should use flexible wire, multistranded, but now with an equivalent cross section of perhaps 20 gauge, or even 22 gauge. If you are working with a long line, or a line with a lot of loudspeakers on it, you may want to go to an equivalent cross section of 18 gauge. But the important thing is to be sure it has adequate insulation.

In such an installation, you may have electrical codes to meet, too. Find out what they are. Remember that, although you have 70 volts between lines, this represents only 35 volts to ground from each side. Codes are usually based on the nominal voltages of alternating current supplies. So the fact that the peak voltages are 1.4 times the rated voltages is already taken care of in the way the code specifies it as a rule.

Where you have high-voltage audio, which a 70-volt line really is, it is good to protect the exposed terminals against accidental shock and static "fluff" build-up. If you leave a 70-volt line system exposed to the air, it will attract dust particles. While you may want an electrostatic air cleaner, we do not recommend using your audio distribution system for that purpose.

Another thing you will have to decide on is the material of which the insulation should be made. Audio is not high frequency; unlike UHF or VHF, it does not need a low-loss dielectric. So you do not need insulation with high-performance capabilities in the electrical sense. But you do need a capability of withstanding sharp wave-front voltages.

Cable intended for 220-volt insulation in a line supply situation should be adequate—better yet, for a 400-volt system. The only difficulty you may encounter here is that most of this kind of cable is intended for really high power, which is why voltages higher than 117 are used in the first place, while for your audio, 20 gauge will be heavy enough for equivalent cross section.

However, if you cannot get the insulation you need in such a light-gauge wire—and you may find it difficult—perhaps you can at least save on cost by finding a surplus supply store that will let you have the quantity you need at a reasonable price. The only disadvantage to the unnecessarily heavy gauge for this purpose is its cost, although if it is too heavy, you may find difficulty making connections to the terminals.

The main thing is to be aware of the various problems, then select the best compromise you can from what is available.

Glossary

Absorption. The property of being able to abosrb rather than to reflect sound waves. It is used to compute the acoustic properties of rooms. The standard of absorption is that an open window absorbs sound 100% because all the sound reaching the area of wall occupied by an open window goes out the window. If half as much sound would go out the window is abosrbed by a corresponding area of wall, the wall is said to have 50% absorption.

Acoustic Dipole. A name given to an open-backed loudspeaker, particularly if it is mounted on a flat baffle. Because sound waves originate from both sides of the cone, or diaphragm, but are out of phase, one pushing when the other is pulling, the kind of waves it radiates have similarities with the waves radiated in radio by a dipole antenna, thus the name.

Acoustic Filter. When the internal construction of a loudspeaker cabinet or box is designed to behave like an electrical filter would on the frequency response of the system, but the effect is achieved by working on the acoustic waves generated inside he loudspeaker, the part of the loudspeaker that does this is called an acoustic filter.

Acoustic Lens. The essential features of an acoustic lens are that it is built of channels whose width is much smaller than the highest frequency's wavelength, and that those channels make the sound travel further than they would in free air without the channels. The extra time taken traversing this extra distance has the effect of forming a sound wave, just as an optical lens forms the waves propagated by light.

211

Acoustic Loading—Back-Loaded Horn

Acoustic Loading. When the radiation of a sound wave produces loading on a loudspeaker diaphragm, that part of the energy used to drive the diaphragm is called acoustic loading. More particularly, it is applied where the arrangement of the system increases that loading on the diaphragm, as when multiple units produce such loading on one another, or where restrictuion, as in the throat of a horn, increases the loading.

Acoustic Suspension. A word used to describe a system designed so that the main controlling force on the diaphragm is the compression and expansion of air inside a sealed loudspeaker box.

Acoustical Transparency. Many light, open-weave fabrics, or types of expanded metal or plastic, that allow sound to pass through virtually unimpeded, although they ar not optically transparent. Such substances are called acoustically transparent.

Ambience. Primarily the background sound present in a room in which you listen to program sound. It can include extraneous sounds, such as audience noises, reverberation of the programs sound, and any other components that would normally be present in background sound.

Anechoic Room. A specially constructed room in which all of the walls, including ceiling and floor, are made highly absorbent to sound throughout most of the audible range, so that there are no reflections or echoes. It is not a natural listening environment, but is useful for conducting measurements on loudspeakers or microphones.

Antiresonance. The opposite of a resonance. An antiresonance is a frequency at which a system absorbs, or fails to radiate, the normal amount of sound.

Attack. Part of a program sound that begins with a sudden component that is louder than the rest of the sound. Applied to a system, the capability of satisfactorily reproducing or rendering such a sound.

Audio Salon. A showroom designed for prospective customers to audition loudspeakers and other system components before purchase. In using one, you need to verify that its acoustic properties—size and treatment—are as close to being like the room for which you want the components as possible.

Back-Loaded Horn. A type of folded horn in which the back of the loudspeaker diaphragm or cone feeds into a horn designed to handle the lower frequencies, while the front of the diaphragm radiates the middle and upper frequencies, either as a direct radiator or with a smaller front-loaded horn.

Baffle. Primarily any device designed to prevent air movement from the front of a loudspeaker cone or diaphragm from having easy access to the rear. In its simplest form, it is a flat board with a hole in it of a size suitable to accommodate the loudspeaker unit for which it will be used. See *Infinite Baffle*.

Bass. The lowest frequencies that a sound system is called upon to handle. In music it will be the major frequencies produced by instruments classified as bass instruments, as opposed to tenor, alto, or soprano. In sound reproduction it will be the frequencies handled by the biggest unit of a multiway system. In a single-unit system, it applies to the lowest frequencies the system reproduces.

Bass Reflex. A loudspeaker system in which the lowest frequencies reproduced are reinforced by an acoustic combination that phase-reverses pressure from the back of the diaphragm so that it emerges in phase with sound waves from the front.

Beaming Sound. Any of a variety of methods of concentrating the direction of sound into wanted areas, where an audience is or may be listening, while restricting it from radiating into areas where excessive reflection could cause confusion or excessive reverberation.

Biamplification or Triamplification. A system using two-way or multiway loudspeakers, in which separation of the frequencies is achieved before final power amplification so that each unit of the system is driven by a separate power amplifier.

Binaural Listening. A property about the way any person with two normal ears listens, enabling him to readily determine the direction from which sounds reach him. The human hearing faculty processes the combined sounds received by each ear to determine time and intensity differences, from which information about direction of the arriving sound is conveyed to the listener.

Binaural Recording. A form of recording that uses two channels of sound, a little differently from stereo. It is designed primarily to be used with headphones so that the binaural listening faculty gives an accurate reproduction of the illusion that would have been conveyed to the listener had he been present at the original performance.

Bookshelf Speaker. One of a variety of loudspeaker designs, in which more than the usual bass response is obtained from a much smaller unit, of a size suitable for placing on a bookshelf.

Buzz. A sound, usually generated mechanically, that is spurious to the program sound. A typical source of such sound would be the diaphragm tapping against the grill cloth on high-amplitude bass response. It could also be due to vibration effects producing spurious sounds in parts of the cabinet that are not solidly secured. Occasionally, a similar sound may be produced in the electronic part of the system by a form of intermodulation.

Classic Stereo. In the early days of stereo, to get the correct illusion the listener was required to sit at an equal distance from the two loudspeakers reproducing left- and right-channel sound. Now stereo for which that would be a requirement, or seating oneself in such a position, is referred to as classic stereo.

Closed Pipe. Part of an acoustic system built by analogy with a closed, or stopped, organ pipe. It is characterized by providing emphasis of frequencies such that their wavelengths make the closed pipe an odd number of quarter-wavelengths long.

Coloration. A property of a loudspeaker, or other part of a system, that changes the relative intensity and response to certain groups of frequencies in such a way as to give the sound a different character, or color. The objective, in any good reproduction, should be to make the reproduced sound audibly indistinguishable from the original sound. Usually, the telltale differences constitute a form of coloration.

Column Speaker. A composite loudspeaker system made up of a line of units all working in unison with the result that sound is concentrated in a cylindrical pattern, of which the line of column is the axis. Column speakers are a convenient way of beaming sound.

Compatability. Whenever a new system is introduced, such as stereo in the days of mono, or quadraphonic in the days of stereo, one of the problems encountered is that a compatibility: old systems should be able to reproduce a new program acceptably, and new systems should be able to reproduce old program acceptably, each in addition to the new system being able to do its best with the new program.

Compliance. Another word for elasticity. It applies to the restoring force used to bring the diaphragm back to its normal position after reproduction of a sound wave has moved it away from that position. The lower the restoring force, the greater the compliance. An everyday word that conveys the same intent is springiness.

Constant-Voltage Distribution. A system in which the voltage delivered to all loudspeaker units in the system is the same, and the relative power taken by each is controlled by changing the impedance presented to the system by that individual unit. Constant voltage refers to the maximum or peak voltage reached when the system is momentarily delivering full power.

Corner Horn. A type of horn loudspeaker unit designed to be placed in the corner of a room, and in which the two walls and floor that constitute the corner provide the completion of the horn for the purposes of getting the lowest frequencies out into the room.

Crossovers. Electrical networks or filters, the purpose of which is to see that the correct frequencies get delivered to the various units of a multiway system.

Crossover Frequency. The precise frequency in a crossover at which each unit receives equal power. At lower frequencies most of the power goes to one unit, and at higher frequencies most of the power goes to the other unit.

Cylindrical Wave. A form of sound radiation in which sound moves out like an expanding cylinder. If you think of a can, or cylinder, very little sound goes out along the axis of the cylinder or can, most of it moves outward from the curved surface.

Damping Factor. A property of amplifiers designed to stop spurious movement of loudspeaker diaphragms when reproducing complex sound waves that include transients. It does so by providing a braking force on the diaphragm whenever the active driving force suddenly terminates. If the internal resistance of the amplifier is one-tenth of the nominal loudspeaker impedance to receive rated power, the amplifier is said to have a damping factor of ten.

dB levels. A system of rated levels of sound with reference to the threshold of audibility or hearing. See *Decibel*.

Dead Room. A room with more than average absorbent surfaces so that sound does not tend to bounce around or reverberate in it.

Decible. A unit of loudness difference. Changes in apparent loudness depend on the ratio or factor by which the sound energy producing them is changed. Thus increasing sound energy by ten times produces a loudness difference of 10 decibels. A decibel is a barely discernible loudness difference, such that ten of them represent an increase of sound energy by ten times.

Diaphragm. The moving part of a loudspeaker unit, usually conical or similar in form, that is responsible for producing the air movements that eventually form a sound wave.

Diffusion. The capability of a loudspeaker to send sound waves out relatively uniformly in all directions.

Dipole. See *Acoustic Dipole*.

Directive Listening. The human capability of concentrating attention on sounds coming from a specific direction. As the listener has no power to change the sound waves reaching his head, and thus his ears, the power of directive listening is totally within the auditory interpretive faculty of a person's brain.

Directivity, Loudspeaker. The capability of a loudspeaker to concentrate sound radiated to certain directions and to the exclusion or reduction of others.

Direct Radiator. Any loudspeaker in which the diaphragm radiates sound waves directly into the surrounding air rather than, for example, through a horn.

Displacement, Acoustic. The total movement of air accompanying the radiation of a given sound wave. Sometimes also called volume displacement.

Displacement, Mechanical. The total movement accompanying the radiation of a given sound wave, for example, of the diaphragm. It would thus be the total distance moved by the diaphragm when radiating a sound wave of given frequency and intensity.

Drone Cone. In some of the reflex units, a cone mounted in an opening that is not electrically driven, but free to move as part of the radiation system for some of the lowest frequencies the system handles.

Duct. Part of a loudspeaker enclosure, usually in the form of a tube, circular or rectangular in shape, through which air moves or escapes, particularly at the lowest frequency the loudspeaker unit handles. See also *Port* and *Vent*.

Efficiency. The amount of acoustic energy radiated as a sound wave, expressed as a fraction, or percentage, of the electrical energy required to produce it. Loudspeakers are not inherently high-efficiency devices, by their very nature. The highest efficiency systems ever built were a little over 50% efficient. What is commonly called a high-efficiency unit may be between 10% and 20% efficient. Low-efficiency units have efficiencies well below 1%.

Electronic Crossover. A crossover for use with biamplification. Instead of using reactances, such as inductors and capacitors, it uses active devices, such as transistors, to achieve the necessary frequency response. It should be noted that the combination of amplitude and phase response produced by an electronic crossover is identical with that of a prototype crossover of the nonelectronic type. There are other misconceptions about their use discussed in preceding chapters.

Enclosure. A general word for the box in which a loudspeaker unit is installed.

Equivalent Duct Length. The effect of a duct in a loudspeaker system that is due to the air column that moves through it. As the air in motion includes that approaching and leaving at the ends of the tube itself, this means the equivalent length is always longer than its physical measurements.

Exponential Horn. A horn in which the expansion rate, or flare, is such that its cross-sectional area doubles at a regular distance interval along its length. It has a cutoff frequency dependent on this flare rate and the mouth size, reproducing frequencies above the cutoff frequency but not below it.

Extension Speaker. A loudspeaker installed in a different room or location from the main loudspeaker. It may be for use at the same time as the main loudspeakers, or separately. Connecting it for best use depends on just how it will be used, requiring care in figuring how to do it.

Faders. In professional systems, relative volume levels from different sound sources are controlled by faders. Originally intended for programs where such sources could be faded in and out, these controls are really well-designed step switches, giving a change in level of about 1 dB for each step. A typical fader may have from 30 to 60 steps.

Flare Rate. The rate at which an exponential horn expands the sound wave. The professional formula uses a coefficient in the exponent of ϵ to represent flare rate. For quick and easy calculations, a good substitute is the distance along the horn at which area doubles. In inches this can be taken as 700 divided by the horn cutoff frequency. Thus for a 100 hertz cutoff, the area doubles every 7 inches.

Flat Radiator. A type of loudspeaker using a slim-line unit mounted in a flat baffle, and used as an acoustic dipole. For certain types of rooms, a pair of flat radiators can give very good stereo reproduction.

Flush Mounting. As with anything, flush mounting means mounting the article so its front edge is flush with the surface in which it is mounted. However, in loudspeaker installation, flush mounting requires careful attention to see that it is acoustically flush so that no cavities produce spurious effects not desired.

FM Multiplex. A system for transmitting two channels of stereo or, by further adaptation, four channels of quadraphonic, over a single FM carrier. In essence it uses a 38 kilohertz switching frequency to carry alternate samples of left and right program. The 38 kilohertz frequency itself is not transmitted, but a 19 kilohertz pilot frequency is transmitted to provide a means of

Folded Horn—Horn

synchronizing the switching at the receiver to coincide with that at the transmitter.

Folded Horn. A way of reducing the space occupied by a horn. The usual way turns the horn back on itself so the sound path divides into multiple channels, or an expanding channel, covering distance down the horn without requiring the same physical distance from throat to mouth. See *Horn* and *Back-Loaded Horn*.

Frequency. The rate of vibration that is responsible for the characteristic of sound known as pitch. The more rapidly the vibrations occur, thus the higher the frequency, the higher the pitch of the sound heard. Human hearing is sensitive to a range of frequencies from approximately 20 hertz, or vibrations per second, to approaching 20,000 hertz.

Hard Surface. A term used in acoustics to indicate a high degree of sound reflection. A surface that is hard acoustically may or may not be very hard in the physical sense because air, the medium through which sound normally travels, is very soft by comparison.

Harmonics. The pitch of a note is determined by its fundamental frequency. Most sound sources produce multiples of the fundamental frequency, which are called harmonics. The second harmonic (which in music is known as the first overtone) is twice the fundamental frequency, the third three times, and so on. The term harmonics is used by audio engineers, while musicians prefer overtones.

Hearing Faculty. While human ears are a vital part of a person's hearing faculty, what we hear is determined more by the processing center in the brain. This compares the nerve impulses corresponding to sounds received by the two ears, and from them conveys to our consciousness, within limits, whatever we want to hear.

Hole in the Middle. In a stereo system, when the left and right loudspeakers are too widely separated for the room in which they are installed, sounds that should appear to come from center lose their identity, and an impression is conveyed that there is a "hole in the middle." Some systems overcome this by providing a mix between left and right to obtain a center channel, which is fed to a center loudspeaker. Modern stereo techniques have rendered this unnecessary in most instances.

Horn. Before the advent of loudspeakers, the musical instrument family known as horns produced families of notes, based on resonant properties of the horns. For a loudspeaker, a special development of the shape is intended to amplify all frequencies within the horn's range uniformly (see *Exponential Horn*). Other special developments are designed to distribute the sound in special ways (see *Multicellular Horn*).

Impedance. The property about a loudspeaker that determines the ratio of voltage to current that it draws from the amplifier to produce sound. Impedance is given a nominal value, such as 8 ohms. However, impedance varies at different frequencies, with consequences discussed more fully in preceding chapters.

Infinite Baffle. A baffle is more effective, the larger it is. An infinite baffle totally encloses the back of the loudspeaker cone or diaphragm in a sealed box so there is no acoustic path from the front to the back of the diaphragm, thus the baffle is effectively infinite in size. It does, however, have other limitations.

Integrated Design. A design in which the enclosure and the unit to work in it are designed to go together, such as the acoustic suspension or loaded reflex types. With such designs, neither will work correctly unless the other is used with it.

Intermodulation. A form of distortion that occurs only when more than one frequency is present in the reproduced sound. It occurs due to interaction between two or more frequencies. Two main forms are distinguishable. In one a relatively low frequency modulates much higher frequencies, causing a sound resembling gargling, as if the higher frequencies come through a "throat" being gargled at the lower frequency. In the other, two higher frequencies produce a spurious note of lower frequency that resembles a buzz.

Intimacy. A property about program sound that conveys the impression of a small, intimate group, of which the listener is a part. It should be marked by a relatively low reverberation and a frequency balance that suggests closeness, rather than distance.

Lens, Acoustic. See *Acoustic Lens*.

Listening. Paying auditory attention. This involves not only sound reaching the ears of a listener, but the listener's using his hearing faculty to critically examine the sounds heard for whatever purpose. With speech, this could be the gathering of information. With music, it could be solely for enjoyment.

Live Room. A term used to describe a room, in which most surfaces reflect more sound than they absorb. From the viewpoint of installing effective loudspeaker systems, the live room presents more problems than the dead room because ways must be found to use its liveness, rather than have the liveness create confusion in sound.

Loaded Reflex—Mono

Loaded Reflex. A variety of modifications to the bass reflex that extend its low-frequency capability, at the expense of efficiency, in a manner similar to the way acoustic suspension extends the capability of the infinite baffle.

Loudspeaker Directivity. Any difference in the way a loudspeaker delivers sound in different directions. This may be a situation in which some frequencies (usually the highs) are delivered more directly than others (the lows), or it may be a situation in which careful design has enabled all or most frequencies to be delivered in similar directional fashion.

Lumped Parameters. Used as a description of any system in which different parts of the system utilize the parameters of the air—its compressibility and mass—separately, as opposed to a system in which the propagation velocity of sound is an essential ingredient in the design. Notably, in a system that uses lumped parameters, the important feature of air inside a box is its compressibility, not the time take for sound to travel the distance across it; the important feature of air in a duct is its mass, rather than the time taken for sound to traverse its length.

Masking. A situation in which a louder sound prevents you from hearing a quieter one. Because the quieter sound is not heard, the listener is usually not conscious that masking occurs. For this reason the effect can have a great many results that the average listener does not suspect. Possibly the most noteworthy is the change in apparent threshold of hearing, or audibility. However it can also affect apparent loudness of sounds well within audible limits under varying ambient conditions.

Mass. In the case of a loudspeaker diaphragm, or cone, mass is simply its weight, usually measured in grams. However acoustic mass is more complex. Primarily it refers to the density of air in weight per unit volume. But in loudspeaker usage, it can refer to the mass effects in a port or duct, which vary in rather complicated fashion with the dimensions of the opening.

Microphone. An instrument for picking up sound. While beyond the scope of this book, microphone techniques are inseparably connected with the problems of reproduced or reinforced sound. Microphones have similarities and differences from those properties discussed in this book relative to loudspeakers.

Midrange. Frequencies between the low, usually handled by the woofer, and the high, usually handled by the tweeter. The name given to the unit that handles these frequencies.

Mono. Short for monophonic. Sound from a single channel or source. At one time called monaural, this term was discarded because it does not mean the listener hears with only one ear. While for many purposes stereo or quadraphonic reproduction provides more satisfaction, mono still has its uses, especially for aligning the more complicated systems.

Mouth. The end of a horn loudspeaker from which sound emerges into the air. Its dimensions are important for determining the low-frequency cutoff of the horn. A useful formula is that the dimension across the mouth needs to be at least 4000 inches, divided by the cutoff frequency in hertz. Thus a horn with 100 hertz cutoff should have a mouth at least 40 inches across.

Multicellular Horns. A type of horn in which the throat divides the sound channel into a number of separate channels, each of which feeds its own individual horn. These are usually arranged in clusters to provide solid coverage for the frequencies delivered by the horns, over an appropriate angle, proportionate to the number of horns into which that angle is divided.

Multiple Unit Systems. A type of loudspeaker (of which the column speaker is an example) in which a number of small units are used, instead of a single larger one. Such a system has the advantage that it can handle lower frequencies, somewhat after the manner of a single larger unit, while at the same time handling higher frequencies, the same as each individual unit in the multiple unit system can.

Multiway Systems. A type of loudspeaker employing some or all of the following elements: a woofer, tweeter, a midrange, a superwoofer, a supertweeter, and suitable crossovers or biamplification to feed them all correctly.

Music. From the viewpoint of sound reproduction or reinforcement, music is defined as sound in which individual components heard are identified primarily by their frequencies as having various differences in pitch. While, esthetically, one man's music is another man's noise, from the viewpoint of loudspeaker installation, this is the important distinction from sounds generally classified as voice.

Music Power Rating. A method of rating the output capability of an amplifier. Because peaks in music are often of relatively short duration, so that the amplifier is not called upon to deliver such rated power continuously, the music power rating is intended to be the maximum power that the amplifier would be required to deliver if the peak music power were sustained, instead of existing only for instant now and then during the program material.

Nonlinearity. A general expression to mean a variety of defects in performance. Nonlinearity in frequency response means that the system does not reproduce all frequencies uniformly. More generally, nonlinearity means that the cone or diaphragm movement does not uniformly correspond to the driving force, or the electrical output from the amplifier. The consequence of nonlinearity is that the system produces harmonic distortion, the introduction of harmonics not present in the original sound, and intermodulation.

Octave. The basic unit of the musical scale, that interrelates frequency and pitch. In frequency, each octave doubles or halves frequency. In pitch, an octave interval is twelve semitones, and one that provides the only perfect unison when two notes are sounded together.

Omnidirectional Speakers. Literally, any type of loudspeaker that delivers sound uniformly in all directions. More specifically, a type that delivers sound uniformly in all horizontal directions. Most units so designated consist of units mounted facing vertically upwards in a pillar type enclosure with a shaped piece over the unit to assist in dispersing the sound horizontally in all directions.

Outdoor Effect. Sounds produced indoors are accompanied by much more reverberation than is characteristic of sounds produced outdoors, especially if the indoor space is large, such as a sound stage. Outdoor effect consists of using microphone techniques and other processing to give the illusion that the sound is outdoors. Even more specifically, low-frequency sounds such as a foghorn can have directionality outdoors, which it is virtually impossible to reproduce indoors.

Overtones. The musicians' name for what loudspeaker designers and electronic engineers call harmonics. However, the second harmonic, which is twice the fundamental frequency, is called the first overtone, the third harmonic becomes the second overtone, and so on. The first harmonic becomes the second overtone, and so on. The first harmonic is the fundamental.

Parallel Connection. A method of connecting loudspeaker units, crossover filter inputs, or whatever, so that all of the items so connected share the same voltage, and distribute the current between them. Thus, while they all receive the same voltage, the current into all of them is the sum total of the individual currents into each of them. Parallel connection is most advantageous for applying an amplifier's damping factor to every unit in the system.

Particle Velocity. The maximum velocity at which particles of air move during the passage of a particular sound wave due to the passage of that sound wave. This should not be confused with propagation velocity. For a single frequency of sound, particle velocity reaches alternate maxima in each direction, normally along the direction of propagation, although in the case of flat radiators or dipoles particle velocity has components transverse to the direction of propagation, and may even follow circular or elliptical paths.

Peak Power. The maximum momentary power represented, during reproduction of sound, roughly corresponding to a moment of maximum compression or expansion of air in the waves produced by the loudspeaker. In a single frequency sine wave, peak power is always just twice the average power. In any program sound, peak power is much more than twice average power, often as much as 10 or 100 times average power.

Phase. A way of representing the different parts of a wave, or of a combination of waves, where a designated angle describes the relative timing. Such a phase angle must always has a reference, or point in relative time corresponding to the wave being measured, to which other points are compared. A complete cycle, or period of vibration is always 360 degrees, and parts of a cycle, or wave, are designated proportionately.

Phasing. Attention to the way various units in a system are connected so that the correct combination of waves shall be radiated at all frequencies by the whole system. This can apply to different units within a multiway system, or a multiple unit system, to different loudspeakers on a monophonic system, to loudspeakers on a stereo or quadraphonic system, and to combinations of all these. In order to phase a whole system other than monophonic, it is best to use monophonic program to avoid confusion due to phase differences between channels of a stereo or quad system.

Ping-Pong Effect. The use of a stereo system and its program to produce extreme separation by having first a sound reproduced from the left channel, then the right, then back to the left, and so on, with virtually no inbetween. While it is a dramatic demonstration of stereo, it does not achieve the advantage for which stereo is capable and has now been discontinued by most producers of programs.

Pipe, Open or Closed. Unlike a duct, in which air acts mainly by its mass, the use of an acoustic pipe is based on organ pipe technology. A pipe open at one end, closed at the other, emphasizes resonances at frequencies for which the pipe is an odd number of quarter-wavelengths long. A pipe open at both ends emphasizes resonances at frequencies for which the pipe is any number of half-wavelengths long.

Pitch. The quality of musical sounds that corresponds to frequency. See *Octave* and *Semitone*.

Plane Wave. A sound wave that travels forward without expanding appreciably. A small section of a sound wave at a considerable distance from its source always approximates a plane wave although, strictly, it may be part of a spherical or cylindrical wave. Alternatively, a plane wave may be created deliberately by use of a special kind of sound source.

Port. In a bass reflex or similar enclosure, an opening that does not have a duct attached. See *Duct* and *Vent*.

Power. Energy being produced or reproduced in any form. It is measured in watts. However, energy and power are measured over time, whereas in the production and propagation of sound waves, these quantities are changing at a very rapid rate. This leads to the need for specifying such quantities as average power, peak power, maximum power, and so forth.

Power Distribution. A system of connections to enable the power delivered by an amplifier to be distributed properly to the various loudspeaker units it has to drive so that sound is properly distributed.

Pressure, Sound. The maximum variation in pressure of air from its average (barometric) value due to the passage of a sound wave. Maximum sound levels may produce measurable variation from barometric pressure. However, sound levels closer to the limit of audibility (threshold of hearing) have sound pressures that are about one-millionth of maximum, so special means are needed to detect and measure them. Sound power is the product of multiplying sound pressure and particle velocity, both of which, in a normal sound wave, vary proportionately.

Propagation. The process by which a sound wave propels itself forward. Momentary sound pressure and momentary particle velocity, each mutually produce the other, which results in the propagation of sound waves.

Propagation Velocity. The characteristic velocity at which sound waves propagate. In air, this is about 1080 feet per second. In water, and in most solids or liquids, it is at least five times as fast as that. Propagation velocity depends on the density (or mass per unit volume) and elasticity (or compressibility) of the medium through which the sound is traveling.

Psychoacoustics. The study of the effect that sounds people hear have on them. Relative to sound reproduction and reinforcement, it pertains more specifically to determining what they can hear and to how they tell differences between various sounds, according to the source of such sounds and the environment in which they are heard.

Quad. Short for quadraphonic. Any of a variety of systems, in which four channels of sound are presented with the aid of at least four loudspeakers in order to enhance the realism of sound reproduction. In most quad systems, two such channels handle essentially direct sound, coming from the original sound sources, while the other two reproduce ambience from the original sound with the purpose of reproducing the original environment.

Reactance. One of two kinds of elements used in crossovers: inductance and capacitance. The essential feature of any reactance is that its value changes with frequency, enabling a crossover built of such reactances to deliver different frequencies presented to its input to different outputs, to which different loudspeakers, such as woofers, and tweeters, are connected.

Reflex, Bass. See *Bass Reflex*.

Reflex, Loaded. See *Loaded Reflex*.

Reinforcement, Sound. A system in which the purpose of a loudspeaker system is to make sound more audible in parts of a room or auditorium where the original sound is produced. This introduces quite different requirements for the system than for sound reproduction, mainly because the original sound and the reinforced sound have the possibility of interfering with one another.

Reproduction, Sound. A system in which the purpose of a loudspeaker system is to reproduce sound in a location different in place and time, or both, from that where the original sound was produced. The characteristic difference from sound reinforcement is that there is no direct connection, acoustically, between the original sound and the reproduced sound.

Resonance. A property of a system that emphasizes one particular frequency, or a narrow band of frequencies. Because it depends on a number of elements or factors, the frequency at which it occurs can vary when changes are made in the system. Thus a loudspeaker unit resonance may change considerably, according to whether it is mounted or unmounted, as well as with how it is mounted.

Resonator. A combination of elements to produce resonance. The best known example of a resonator is the Helmholtz resonator, in which the elements are the air inside a bottle with the air in a comparatively narrow neck that the bottle has.

Reverberation. In one sense, a fancy word for echo. However, where echo usually refers to a single, and quite distinct repetition of the original sound, reverberation more generally refers to a much more confused multiplicity of such repetitions, sometimes with insufficient time difference to be discernible as a separate entity, although their presence changes the nature of the sound heard.

Reverberation Time. This is one measure of reverberation, being the time taken for sound level to drop by 60 dB after a source of continuous sound (to fill the room with reverberant sound) is abruptly discontinued. But this simple measure of reverberation time gives only a very rough idea of how much reverberation a room or auditorium possesses, giving virtually no idea at all about its character.

Sealing. A very necessary procedure with any loudspeaker enclosure that is enclosed at the back to prevent leaks and spurious sounds.

Semitone. A unit on the Western musical scale, being one-twelfth of an octave. The ratio of one semitone frequency to the next is 1.059463:1. To illustrate what this means, suppose we start from a note of 1000 hertz (this is not a musical tone in any standard scale): the frequencies from that note up one whole octave would be: 1000, 1059.63, 1122.46, 1189.21, 1259.92, 1334.84, 1414.21, 1498.31, 1587.40, 1681.79, 1781.80, 1887.75, 2000.

Separation, Stereo. This can have two meanings: (1) a performance spec, which might say that separation is, say 40 dB. This would mean that, if some program is supposed to be 100% from the left channel, only 1% leaks through to the right channel; (2) an indication of how well the separation sounds to an average listener.

Series Connection. A method of connecting loudspeaker units, crossover inputs, or whatever, so that they receive the same input current. The total voltage is shared between them, so they each get a part of it. Thus they each get a part of the total voltage, while sharing the same current. This method of connection may be the best available in multiple unit operation, but it is not recommended if parallel can be used.

Soft Surface. A surface that has good absorbency for acoustic or sound waves. As with hard surfaces, this is not necessarily related to its physical characteristics otherwise. For example, a well-designed acoustic tile may have an acoustically soft surface, while being quite hard physically.

Sound Pressure. See *Pressure, Sound.*

Sound Reinforcement and Reproduction. See *Reinforcement* and *Reproduction.*

Spherical Wave. The most basic form of acoustic wave that radiates outward from a single, central source. So called, because each expanding shell of instantaneously equal sound pressure is spherical in form. While this is the most basic form, a perfect spherical wave is actually rare, only approximated in real life.

Standing Waves. A form of wave that does not appear to move. To produce it requires a steady source of sound at the frequency of the wave in a room with reflecting walls such that reflections cause a steady pattern to build up with nodes and antinodes of sound pressure all over the room. A pipe produces the simplest form of standing wave in a single direction along the pipe.

Stereo. Short for stereophonic sound, which means literally "solid sound." By using two or more reproducing sources (loudspeakers) stereo can create an illusion of sound sources distributed about the room, other than where the loudspeakers are located, that to some extent reproduces the positions of the original sound sources as picked up by the microphones.

Stereo Record. A variety of phonograph record in which the groove, and thus the stylus that follows it, moves in a variety of directions as the record rotates. The easiest way to visualize how a single groove can thus carry two channels of program sound is to think of the left channel as being recorded by undulations on the left wall of the groove, while the right channel is recorded by undulations on the right wall of the groove. Program representing sound at front center, comes from a groove in which the groove moves purely from left to right, with no vertical motion at all, so that both walls of the groove have equal undulations on them.

Stereo Separation. See *Separation, Stereo.*

Successive Decoupling. Design of a loudspeaker cone or diaphragm, so that at progressively higher frequencies, less of the cone moves. At the lowest frequencies the whole cone moves, flexing at the outer surround. As frequency moves up, outer rings of the cone gradually stop moving, as parts of the cone itself start to flex, allowing the inner rings to move at these frequqncies without the outer ones. Properly designed, this makes for a more uniform frequency response.

Supertweeters. Extra precision, tiny units designed to reproduce the very high frequencies, above those where ordinary tweeters start to become somewhat erratic in their response.

Superwoofer. An extra large, heavy-duty woofer, designed to reproduce only the very lowest frequencies. A secondary purpose for such a unit, in some instances, is to reproduce airborne vibrations that are too low to be considered audible sound. For example, to produce effects for a movie such as *Earthquake*.

Suspension. The part of a loudspeaker that provides the restoring force to keep the voice coil and cone in its normal position from which it should move to reproduce sound symmetrically. Parts of a loudspeaker suspension consist of the spider that holds the voice coil centered in the magnetic air gap, the corrugated surround that allows the cone to flex at its outside edge, and the air cushion behind the cone, in the case of a closed box such as an infinite baffle or acoustic suspension type.

Sustain. An effect where a tone dies away slowly instead of abruptly. This is a feature that is often built into an electronic organ, but is usually an undesirable feature in a loudspeaker intended for sound reproduction or reinforcement.

Threshold of Hearing. A sound level that is barely audible to the average listener against a background of complete silence. If any ambient sound is present, the apparent threshold of hearing is raised in level. However, the average true threshold of hearing (the average taken of a number of people with normal hearing) is used as the reference level, against which various sound levels are measured.

Throat. The end of a horn to which the diaphragm of a compression driver or other loudspeaker unit attaches. Its area is smaller than the diaphragm that drives it to provide matching between the mass of the diaphragm and the column of air in the throat of the horn.

Timbre. A musician's word for the character of a musical sound. It is related to its overtone or harmonic structure or makeup the way the sound builds up and decays and various other properties of the sound that enable the listener to recognize what kind of instrument it is, or what kind of instrument's sound is being reproduced.

Tinning Joints. In making soldered connections in wiring, tinning the joints before they are made is vital. This means making sure that both parts to be connected are thoroughly "wet," or covered with melted solder, so that when the solder is applied to the joint together, it makes a secure joint.

Transients. Applied to reproducing sound, transients means sounds that are changing. While any changing sound is a transient, it particularly refers to sudden changes in sound, such as percussive sounds from drums, cymbals, etc.

Transducer. The part of a loudspeaker, microphone, or any device that transforms program sound from one form to another. In the case of a loudspeaker, the moving coil converts the electrical currents and voltages from the amplifier into drive force and movement of the cone or diaphragm that moves air to form sound waves. Thus the moving-coil element, with its associated magnet system, constitutes the transducer of a loudspeaker.

Transparency, Acoustical. See *Acoustical Transparency*.

Treble. In music, sounds above middle C on the concert-pitch scale, which is approximately 260 hertz. In sound reproduction the word has a somewhat more flexible meaning. Middle C is about the middle of what is usually called midrange, so treble is usually taken to begin at a somewhat higher frequency, such as 500 or 1000 hertz.

Triamplification. A strict word that should be used for a three-way system, where each unit has its own power amplifier. However, the word biamplification is used almost universally, whether the number of ways used is two or more.

Tweeters. Loudspeaker units designed to handle the higher frequencies, treble or higher. The frequency at which a tweeter takes over is determined by the cutoff frequency of the tweeter, and by the crossover that effects the change from woofer or midrange (as the case may be) to the tweeter. An important function of tweeters is to assure smooth or uniform response at these upper frequencies, giving good fidelity to the varying timbres in sound to be reproduced.

Velocity. A fancy word meaning speed of movement. See *Particle Velocity* and *Propagation Velocity*.

Vent. A general word to cover secondary openings in boxes used to produce a bass reflex design, or a loaded reflex design. See *Port* and *Duct*.

Voice. One time more commonly called speech. However, to distinguish from music forms of program, the word voice is preferred. For example in two-way communication systems where the people talking change the direction of communication, the device that activates the change is called a voice-operated relay. Speech more accurately describes what the program consists of, while voice describes the characteristics that distinguish it, as in voice prints.

Volume. At one time, this word was used to describe what is now more commonly called loudness. In the design of loudspeakers, volume applies to the interior content, or space, which is usually calculated in cubic inches or cubic feet as appropriate.

Watts. The unit used to measure power. In an electrical system, such as an amplifier, watts are measured as volts times current in amps. Thus an amplifier that delivers 32 volts into an 8-ohm load, will deliver 4 amps, or 128 watts of power. Watts are used for peak power, average power, maximum power, or any other measure, merely specifying what form of power measurement it is. Acoustic power may also be given in watts by using the efficiency factor of the loudspeaker that converts electrical energy to acoustic energy.

Wavelength. The length of a wave, as propagated in air space, from one pressure maximum to the next pressure maximum, for example, at an instant in time. Wavelength is important, because a loudspeaker has to produce waves of various frequencies, and the wavelength corresponding to any specific frequency depends on the propagation velocity and the frequency. Thus with propagation velocity at 1080 feet per second, the wavelength corresponding to 40 hertz is $1080 \div 40 = 27$ feet.

Woofers. Loudspeakers designed to handle the lowest frequencies normally reproduced.

Index

Index

A
Acoustic suspension 61, 120
Adjustment, finer 169

B
Baffles 45
Baffle, Infinite 50
Baffles, practicle 50
Bass 86
Bass reflex 54
Binaural 24
Box size 52
Box size, effective 139

C
Cables 208
Cars 142
Checking crossover phase 157
Coil 43
Columns 105
Connections 203
Connections, making 198
Connecting wire 197
Constant voltage line distribution 189
Crossovers 71, 182
Crossover, electrical or electronic 186
Crossover phase 156
Crossover phase, checking 157

D
Damping 166, 172
Deficiencies, stereo 27
Design, duct 136

Designing your own system 123
Differences in & affecting loudspeakers 8
Directional effect 72
Directivity 67
Distribution, constant voltage line 189
Duct design 136

E
Effective box size 139
Efficiency 28, 60
Electrical or electronic crossover 186
Extensions 176

F
Fault tracing 207
Figuring 169
Finer adjustment 169
Flat radiator 62
Frequencies 65
Frequency 15
Frequency differences 57
Fullbeam multiple unit 188

H
Horn design 95
Horns 46, 93
Hearing differences 10

I
Infinite baffle 50
Integral sound 70
Importance of phase 146
Iron 43

233

J
Joints, soldered 199

L
Leads, routing 205
Listening directively 20
Loaded reflex 123
Loaded reflexes 62
Loudspeakers 7
Loudspeakers design 111
Loudspeaker directivity 22
Loudspeakers to match room 139

M
Making connections 198
Matching 161
Mono, stereo, quad 23
Mounting, snug 143
Multispeaker systems 164
Multiway 180
Multiway systems 57

O
Omnidirectional 78

P
Phase 145
   checking for 148
   crossover 156
   importance 146
Phasing 178
   system 150
Physical fit 41
Placement problems 131
Power 28, 175
Power od selective listening 31

Practicle baffles 50

R
Radiator, flat 62
Range 18
Reflex, bass 54
   loaded 123
Reflexes, loaded 62
Relations between stereo & quad 153
Relative size 74
Resonance, effects 51
Resonators 89
Room quality 34
Room size 36
Routing leads 205

S
Sealing 133
Single unit or multiway 70
Snug mounting 143
Soldered joints 199
Sound differences 11
Sound distribution 19
Stereo 25
Stereo & quad, relations between 153
Stereo Deficiencies 27
Suspension, acoustic 61, 120
System phasing 150

T
Tracing, fault 207

W
Wavelength 17
Wire, connecting 197
Wiring 204